Lecture Notes in Networks and Systems

Volume 138

The series "Lecture Notes in Networks and Systems" publishes the latest developments in Networks and Systems—quickly, informally and with high quality. Original research reported in proceedings and post-proceedings represents the core of LNNS.

Volumes published in LNNS embrace all aspects and subfields of, as well as new challenges in, Networks and Systems.

The series contains proceedings and edited volumes in systems and networks, spanning the areas of Cyber-Physical Systems, Autonomous Systems, Sensor Networks, Control Systems, Energy Systems, Automotive Systems, Biological Systems, Vehicular Networking and Connected Vehicles, Aerospace Systems, Automation, Manufacturing, Smart Grids, Nonlinear Systems, Power Systems, Robotics, Social Systems, Economic Systems and other. Of particular value to both the contributors and the readership are the short publication timeframe and the world-wide distribution and exposure which enable both a wide and rapid dissemination of research output.

The series covers the theory, applications, and perspectives on the state of the art and future developments relevant to systems and networks, decision making, control, complex processes and related areas, as embedded in the fields of interdisciplinary and applied sciences, engineering, computer science, physics, economics, social, and life sciences, as well as the paradigms and methodologies behind them.

**** Indexing: The books of this series are submitted to ISI Proceedings, SCOPUS, Google Scholar and Springerlink ****

More information about this series at http://www.springer.com/series/15179

José Guadalupe Flores Muñiz ·
Nataliya Kalashnykova ·
Viacheslav V. Kalashnikov ·
Vladik Kreinovich

Public Interest and Private Enterprize: New Developments

Theoretical Results and Numerical Algorithms

Springer

José Guadalupe Flores Muñiz
Department of Physics and Mathematics
Universidad Autoonoma de Nuevo Léon
San Nicolas de los Garza
Nuevo León, Mexico

Nataliya Kalashnykova
Department of Physics and Mathematics
Universidad Autoonoma de Nuevo Léon
San Nicolas de los Garza
Nuevo León, Mexico

Viacheslav V. Kalashnikov
Department of Systems
and Industrial Engineering
Tecnologico de Monterrey
ITESM/Campus Monterrey
Monterrey, Nuevo León, Mexico

Vladik Kreinovich
Department of Computer Science
University of Texas at El Paso
El Paso, TX, USA

ISSN 2367-3370 ISSN 2367-3389 (electronic)
Lecture Notes in Networks and Systems
ISBN 978-3-030-58348-4 ISBN 978-3-030-58349-1 (eBook)
https://doi.org/10.1007/978-3-030-58349-1

This Springer imprint is published by the registered company Springer Nature Switzerland AG
The registered company address is: Gewerbestrasse 11, 6330 Cham, Switzerland

Contents

Chapter 1
Introduction

1.1 Motivation and Goals

Researchers in the field of mathematical economics have extensively and intensively studied mixed oligopoly models. In contrast to the classical oligopoly, a mixed oligopoly, apart from standard producers who seek to maximize their net profit, usually includes at least one public company trying to optimize another objective function involving indicators of the firm's social responsibility.

For a long time, mixed oligopolies were studied by making use of the classical Cournot-Nash, Hotelling, or Stackelberg models. However, the notion of *conjectural variations equilibrium* (CVE), first introduced by [1, 2], opens another alternative for the agents' reaction to the market challenge, which attracts ever-growing interest on the part of the related researchers. In CVE, competitors behave as follows: each producer selects its most favorable strategy having supposed that every opponent's action is a *conjectural variation function* of its own strategical variation. For example, as [3] states, "Although the firms make their output decisions simultaneously, plan changes are always possible before production begins". In other words, in contrast to the Cournot-Nash approach, here, every company assumes that its choice of the own output volume will affect the reactions of all the other competitors. The consequently arising prediction (or, conjectural variation) function is the central point of conjectural variation decision-making, or the conjectural variations equilibrium (CVE).

As is mentioned in [4, 5], the notion of CVE has been the topic of abundant theoretical discussions (cf., [6]). Notwithstanding, economists have extensively used various forms of CVE to predict the outcome of non-cooperative behavior in many areas of economics. The literature on conjectural variations has focused mainly on two-player games (cf., [4]) because a serious conceptual difficulty arises if the number of agents is greater than two (cf., [4, 7]).

In order to overcome conceptual hurdles arising in many-player games, a new tool was developed in [8]. Namely, instead of imposing very restrictive additional

J. G. Flores Muñiz et al., *Public Interest and Private Enterprize: New Developments*, Lecture Notes in Networks and Systems 138,
https://doi.org/10.1007/978-3-030-58349-1_1

assumptions (like the identity of players in the oligopoly), it was supposed that every player makes conjectures only about the variations of the market-clearing (equilibrium) price as a consequence of (infinitesimal) variations of the same player's output volume. Knowing the opponents' conjectures (the *influence coefficients*), each firm applies a *verification procedure* in order to determine whether its influence coefficient is *consistent* with those of the remaining agents.

For this reason, the first objective of this book was to extend the results from [8] to the case of a semi-mixed duopoly, where the semi-public company strives to maximize a convex combination of the net profit and domestic social surplus.

When studying an oligopoly market within the theoretical framework of mathematical models, in addition to the results corresponding to the existence and uniqueness of the CCVE, it is especially important to compare with the other two most popular equilibrium kinds: the Cournot-Nash equilibrium and the perfect competition equilibrium. Thus, the second objective of this book was to conduct a comparative analysis between these three types of equilibrium (consistent, Cournot-Nash and perfect competition) for the developed semi-mixed duopoly model.

There are other kinds of markets (e.g. financial markets) and even other types of models (e.g. human migration models) which make use of the concept of conjectural variations but the results concerning to the consistent conjectures cannot yet be extended since it is not possible to directly apply the verification procedure (Consistency Criterion) from [8]. Hence, the third objective of this book was to further study the nature of CCVE in order to propose another way (different from the verification procedure proposed by Bulavsky) to find the consistent equilibrium.

1.2 Contribution

Now, we briefly describe the main results presented within each of the three chapters of this book and the important relationships between them.

In Chap. 2 we prove the existence and uniqueness of the CCVE in a semi-mixed duopoly model, where the semi-public company maximizes the convex combination of the net profit and domestic social surplus, given by a parameter $\beta \in (0, 1]$. Furthermore, we study the behavior of the consistent conjectural variations equilibrium, the Cournot-Nash equilibrium and the perfect competition equilibrium, as functions of β. After analyzing and comparing the results, we developed an *optimality criterion* for the parameter β (which we named socialization level) that allows the semi-public company to benefit the population's economy, as well as its own profit. Hence, in Chap. 2, we achieved the first and second objectives.

The main and most important tool used in Chap. 2 to define the Consistency Criterion was a special verification procedure in which each firm can determine whether its conjecture is consistent with those of the remaining agents. If all conjectures are consistent with each other, these conjectures are named consistent and the corresponding equilibrium is called consistent conjectural variations equilibrium. Chapter 2 contains the detailed description of the special verification procedure proposed by Bulavsky.

As mentioned above, there are other mathematical models in which it is not possible to apply the verification procedure directly. Consequently, it is not possible to extend the concept of consistent conjectural variations equilibrium as presented in Chap. 2. Thus, in Chap. 3, we further study the concept of CCVE. In more detail, we considered a classic oligopoly market (i.e. there are only private firms in the market which maximize their own net profit), presenting the results for the consistency of the conjectures. In addition, the classic oligopoly market was modeled in a second way, as a many-person game (named *meta-game*), where the players are the same private firms of the original oligopoly model and their strategies are their conjectural variations. Then, we prove that (under certain assumptions) the Nash equilibrium for the meta-game is tantamount to the consistent conjectural variations equilibrium for the original oligopoly model.

Therefore, within Chap. 3, we established two important results. The first is that we found another way (different from the verification procedure) to find the consistent conjectural variations equilibrium, by defining the consistent conjectures as the Nash-equilibrium for the corresponding meta-game, which will allow us to extend the concept of CCVE to other types of mathematical models (in which it is not possible to apply the verification procedure). Thus, fulfilling our third objective. The second is that the meta-game can also be formulated as a bilevel multi-leader-follower game. Hence, the search for the consistent equilibrium leads us to the search for the solution of a bilevel problem. Then, the next step is to develop efficient numerical algorithms to find the global solution of a bilevel programming problem in which both the upper-level and the lower-level are a many-person game.

The preliminary ideas for such algorithms are presented in Chap. 4, where we study the Tolls Optimization Problem (TOP), which has important similarities with the meta-game since it is modeled as a bilevel programming problem whose lower-level is a many-person game.

The important results obtained in Chap. 4 are as follows. First, we propose a natural extension for the mathematical formulation of the TOP, considering the delay cost for a single unit to travel across a certain road as a linear function depending upon the total amount of drivers traveling along the same road, instead of a constant as it has been considered before. This new assumption is related to the traffic congestion and leads to a bilevel formulation for the TOP where the lower-level is a convex quadratic many-person game. Second, we reduced the equilibrium problem in the lower-level to a simpler convex quadratic programming problem, which guarantees the existence of a global solution that can be found efficiently with an interior-point algorithm. The efficient solution of the lower-level problem in bilevel optimization is especially important since its optimal solution is required in order to evaluate the upper-level functions. And third, taking advantage of the sensitivity analysis for convex quadratic optimization presented in [9], and the filled function method from [10], we present an efficient algorithm to solves the new (extended) formulation of the TOP.

Recall that in our extended formulation for the TOP, the lower-level is given by a convex quadratic equilibrium problem, which is a particular case of the convex equilibrium problems that appear in the lower-level of the bilevel formulation for the meta-game. Thus, the theory and algorithms developed to solve the TOP will be useful in the future to construct an algorithm that can efficiently find the Nash equilibrium for the meta-game, i.e., the consistent conjectural variations equilibrium.

1.3 Literature Review

Many mixed oligopoly models include an agent who maximizes the domestic social surplus (cf., [11–15]). An income-per-worker function replaces the standard net profit objective function in some other publications (cf., [16–19]). Other researchers [20, 21] have studied a third kind of mixed duopoly, in which an exclusive participant aims to maximize a convex combination of its net profit and domestic social surplus. Such a company is addressed as *semi-public*.

In papers [7, 22], the authors extended the ideas of [8] to the mixed duopoly and oligopoly cases, respectively. They defined *exterior equilibrium* as a CVE state with the conjectures fixed in an extrinsic manner. This sort of CVE was proved to exist uniquely, which helped introduce the notion of *interior equilibrium* as the exterior equilibrium with consistent conjectures (influence coefficients). All these instruments, the consistency criteria, consistency verification procedures, and existence theorems for the interior equilibrium were developed and demonstrated in [7, 22].

In the next series of papers, [23, 24], the aforesaid constructions were extended to the case of a semi-mixed duopoly, where similar to [20, 21], the semi-public company maximizes the convex combination of the net profit and domestic social surplus. The results of numerical experiments with a test model of a market of electricity resembling that of [25], both with and without a semi-public producer, showed that the consumer gains more if the semi-public producer follows the CVE strategy as compared to the Nash-Cournot equilibrium. Furthermore, in [24], the authors declared a guess that there must exist such a value of the combination parameter (also interpreted as the public firm's socialization level) that brings up the "equivalence" (in a certain sense) of the consistent conjectural variations equilibrium (CCVE) and the classical Cournot-Nash one. This equivalence permits a socially responsible municipality to diminish (cancel) subsidies paid either to the private company (in order to compensate its losses when following the consistent conjectures) or to the consumers (to reimburse them the higher retail price of the good if both the competing semi-public firm and the private company both are stuck to the Cournot-Nash conjectures).

In Chap. 2, we present mathematically rigorous proofs of the above-mentioned guess. In other words, we establish the existence of the value of the combination coefficient (also known as the semi-public company's socialization level) such that the private producer's profit is the same in the CCVE and Cournot-Nash equilibrium states, which makes the subsidies from the authorities either to the producer or to the consumers unnecessary.

As mentioned before, the concept of the *conjectural variations equilibrium* (CVE) was first proposed by [1, 2] to extend the concept of a solution to a static (Cournot) model (game).

The papers [26, 27] and the monograph [28] introduce and examine a new form of the CVE, in which the conjectural variations (represented via the *influence coefficients* of each agent) were used to bring about a new equilibrium concept distinct from that of Cournot-Nash.

For instance, in [28], the classical oligopoly model was extended to the conjectural oligopoly as follows. Instead of the usual Cournot-Nash assumptions, all producers $i = 1, \ldots, n$, considered the conjectural variations described below:

$$G_i(q_i + \eta_i) = G + \eta_i \omega_i(G, q_i). \tag{1.1}$$

Here, G is the current total quantity of the product cleared in the market, q_i and $q_i + \eta_i$ are, respectively, the present and the expected supplies by the i-th agent, whereas $G_i(q_i + \eta_i)$ is the total cleared market volume *conjectured* by the i-th agent as a response to changing her own supply from q_i to $q_i + \eta_i$. The conjecture function ω_i was referred to as the i-th agent's *influence quotient (coefficient)*. Recall that the usual Cournot-Nash model assumes $\omega_i \equiv 1$ for all $i = 1, \ldots, n$. Under general enough assumptions concerning the properties of the influence coefficients $\omega_i = \omega_i(G, q_i)$, cost functions $f_i(q_i)$, and the inverse demand function (or, price function) $p = p(G)$, new existence and uniqueness results for the conjectural variations equilibrium (CVE) were obtained. This approach was further developed in [29, 30].

An interesting comparison of the Cournot-Nash and Bertrand models is provided in [31]. The Bertrand's model of oligopoly, which considers the perfect competition, assumes that: (1) there is a competition over prices; and (2) production follows the realization of the demand. The authors from [31] demonstrate that both of these assumptions are required. In more detail, they study a two-stage oligopoly game where, first, there is a simultaneous production, and, second, after the production levels are made public, there is a price competition. Under rather mild assumptions about the demand, the authors show in [31] that the unique equilibrium is the Cournot-Nash one. This illustrates that solutions to oligopoly games depend on both the strategic variables employed and the context (game form) in which those variables are employed.

A different example of a two-stage game can be found in [32]. Similar studies were presented in the unpublished manuscript [33] dealing mainly with forward markets but still providing important insight into the consistent conjectural variations equilibrium regarding the many-stage oligopoly model. The authors examine the impact of the conjectures about the players' knowledge on the outcome of the game, where the outcomes are consistent with their conjectures. Then they deduce the similar result obtained by [34] on forward markets but under different assumptions of knowledge from consistent conjectural variations. All the above-mentioned papers deal in various ways with games with conjectural variations equilibrium (CVE).

In [8] the consistent CVE (given by the special verification procedure) is defined by a system of non-linear equations. Exactly the same verification formulas were obtained independently (10 years later) in [25] establishing the existence and uniqueness of consistent conjectural variation equilibrium in an electricity market. However, to do that, the authors of [25] made use of a much more complicated optimal control technique when searching the system's steady states (a similar approach was employed in [35]). Moreover, in [25], the inverse demand function is linear, and the agents' cost functions are quadratic, whereas [8] allows nonlinear and even non-differentiable demand functions as well as arbitrary convex cost functions for the agents.

As shown in [7], in general, the consistent conjectures are *not* the Cournot-Nash conjectures. In other words, at a consistent CVE, each agent i does not use the Cournot-Nash equilibrium concept since it *does not* assume that all other agents are dead set on the (equilibrium) production volumes q_j, $j \neq i$. In [36], a meta-model is introduced, in which *not* the players' production volumes q_i but their *conjectures* ν_i serve as the players' strategies. The remarkable fact was demonstrated: the consistent (for the original oligopoly) conjectures ν_i^* , while in general not being the Cournot-Nash ones for this game, *are* the optimal Cournot-Nash strategies in the above-mentioned meta-model. In other words, if each player i assumes that the rest of the players stick to their consistent CVE conjectures ν_j^*, $j \neq i$, then, its consistent conjecture ν_i^* is optimal for player i, too. This means that the vector of (consistent in the original oligopoly) conjectures $(\nu_1^*, \ldots, \nu_n^*)$, coincides with the classical Cournot-Nash equilibrium in the meta-game. Similar results were claimed (without proof) in [37] but only for quadratic cost functions.

However, since the meta-model allows the agents to select their strategies from $\mathbb{R}^n$, and this isn't a compact set, the existence of the Cournot-Nash equilibrium in the meta-model has to be guaranteed by some extra assumptions (similar difficulties related to the price equilibrium were run into and overcome by [38]). Under those assumptions, the complete equivalence of the consistent CVE in the original oligopoly and the meta-model has been established in Chap. 3.

In Chap. 4, we study one of the problems affecting megapolises, the traffic congestion in highway networks. The said congestion is a direct consequence of all drivers trying to run the "shortest" path. One can think that a simple solution to this problem might be the construction of more roads in the network, which, however, is usually very expensive. Moreover, this often has the opposite effect of increasing traffic congestion even more (cf., the well-known Braess's paradox published first in [39] and then republished in English in [40]). On the other hand, highway toll pricing has proved to be a convenient tool for decreasing traffic congestion since the drivers now minimize not only their travel time but also its cost. The fees charged on the roads help maintain the highway networks in good conditions. The Tolls Optimization Problem (TOP) deals with selecting and assigning the optimal tolls to the toll arcs in the highway graph. The nature of this problem is generically bilevel because of the existence of multiple decision-makers: (at least) one at the upper-level making decisions concerning the tolls selected with aim to maximize its net profit, and the lower-level highway users (followers) each trying to find the best way along its origin-destination (O-D) path that can include both toll and free arcs. As all the drivers (followers) have to share the same resources (highways) with probably limited capacities, and moreover, their (quadratic) transportation costs might involve flows of the other agents, too, then the lower-level program is, in fact, a typical case of the general Nash equilibrium problem. Therefore, the problem of the leader is to find equilibrium among the toll values that provide high revenues while being attractive enough to the users (followers). The Tolls Optimization Problem has attracted the attention of numerous prominent researchers. In this Introduction, we mention only a few publications that have dealt with the problem in question.

Reference [41] was the first who provided a theoretical structure for the decision-makers at both levels, based on integer programming. It also presented a scheme for uniforming similar network design models and the ways of developing network design algorithms.

Reference [42] pointed out that the Network Design Problem deals with the optimal balance of the transportation, the investment and the maintenance costs of the network subject to the congestion, where the network's users behave according to Wardrop's first principle of traffic equilibrium. Also, in [42], the Network Design Problem is described for the first time as a bilevel programming problem.

Reference [43] handled the Network Design Problem by the analysis of another optimization problem which is inverse to the Network Design Problem.

Reference [44] proposed a primal-dual algorithm generating lower and upper bounds for the maximum profit collected from tolls of the highway network.

Reference [45] developed fuzzy-set-theory methods to approximate the solution to the Tolls Optimization Problem.

Reference [46] also treated the Tolls Optimization Problem by means of bilevel optimization, considering the firm in charge of the tolls as the leader and the drivers as the followers. This problem was also studied by [47] for deciding tariffs on cargo trucks running the highways. In this case, the leader is a group of competing companies whose profit is yielded from the tolls they establish, while the (unique) follower is a carrier minimizing its travel expenses.

Another instance of the TOP was examined in [48], where the leader is a highway administrator maximizing the turnover from the tolls on a subset of the arcs of a network with the drivers as the followers seeking for the "fastest" route (taking into account both the running time and the expenses on the tolls) uniting their origin and destination (O-D) points.

The Tolls Optimization Problem is also set as a combinatorial program which implies it belongs to the class of NP-hard problems (cf. [49]). Apart from the already known NP-hardness proofs, [50] obtained new results about the computational complexity of some popular algorithms.

Reference [51] treated the TOP with the premise that the network could be supplied with a subsidy thus permitting the tolls to have unlimited values. In this paper, they develop an algorithm first constructing paths and then forming columns to reveal the optimal values of the tolls for the current path, which then serve as lower bounds. After that, the upper bound for the leader's profit is updated and finally, a diversification step is applied. The authors continued working on this problem in [52] where they apply a tabu search procedure to conclude that their heuristics produced better results than other combinatorial algorithms.

Reference [53] restated the TOP by using of the optimal-value-function technique which works better than the Karush-Kuhn-Tucker optimality conditions. The optimality conditions for this recasting were deduced and other theoretical properties were exploited.

An innovative approach to solving the TOP was proposed in [54]: as the lower-level equilibrium problem was easily reducible to a (large) linear program, the prospective direction of increase for the upper-level objective functions can be determined with the efficient tool of Sensitivity Analysis. This technique allowed the authors [54] to significantly accelerate the solution of the highly nonconvex TOP in comparison to the previous well-known algorithm.

In the works mentioned above, all the bilevel Tolls Optimization Problems treat linear problems at the lower-level. Thus, the main objective of Chap. 4 is to extend the previous formulations and the promising results from [54] by considering *quadratic* problems at the lower-level (the latter reflect more accurately the traffic congestion on the roads). We also present an efficient algorithm to solve this new form of the TOP making again use of a method based on the sensitivity analysis for convex quadratic optimization presented in [9]. This sensitivity analysis, similar to the technique described in [54] helps determine the allowable variations of the tolls values that do not mess up the optimality of a solution.

In addition to this method based on sensitivity analysis, the proposed algorithm also applies a filled function (FF) technique adapted to our case from [10, 55, 56]. This method is quite efficient when a local optimum is found. The filled function procedure allows to either "jump" to another region of the feasible set which could lead to a better local optimum, or conclude that a good approximation of the global optimum has been found.

Finally, to show the efficiency of our proposed heuristic algorithm, its performance is compared with that of other well-known heuristic algorithms from [57] in a series of numerical experiments with several instances for the bilevel formulation of the TOP with a quadratic lower-level.

References

1. Bowley, A.L.: The mathematical groundwork of economics. Soc. Forces **3**(1), 185 (1924)
2. Frisch, R.: Monopole-polypole: La notion de force dans l'économie. Nationaløkonomisk Tidsskrift **71**, 241–259 (1933). In french, translated to english in "O. Bjerkholt (Ed.) (1995): Foundation of Modern Econometrics The Selected Essays of Ragnar Frisch, Vols. I and II, Edward Elgar, Aldershot, UK"
3. Laitner, J.P.: "RATIONAL" duopoly equilibria. Q. J. Econ. **95**(4), 641–662 (1980)
4. Figuières, C., Jean-Marie, A., Quérou, N., Tidball, M.: Theory of Conjectural Variations. World Scientific, Singapore (2004)
5. Giocoli, N.: The escape from conjectural variations: the consistency condition in duopoly theory from bowley to fellner. Camb. J. Econ. **29**(4), 601–618 (2005)

6. Lindh, T.: The inconsistency of consistent conjectures: coming back to Cournot. J. Econ. Behav. Organ. **18**(1), 69–90 (1992)
7. Kalashnikov, V.V., Bulavsky, V.A., Kalashnykova, N.I., Castillo-Pérez, F.J.: Mixed oligopoly with consistent conjectures. Eur. J. Oper. Res. **210**(3), 729–735 (2011)
8. Bulavsky, V.A.: Structure of demand and equilibrium in a model of oligopoly. Econ. Math. Methods (Ekonomika i Matematicheskie Metody) **33**, 112–134 (1997). In Russian
9. Hadigheh, A.G., Romanko, O., Terlaky, T.: Sensitivity analysis in convex quadratic optimization: simultaneous perturbation of the objective and right-hand-side vectors. Algorithm. Oper. Res. **2**, 94–111 (2007)
10. Renpu, G.E.: A filled function method for finding a global minimizer of a function of several variables. Math. Program. **46**(1), 191–204 (1990)
11. Cornes, R.C., Sepahvand, M.: Cournot vs Stackelberg equilibria with a public enterprise and international competition. In: Discussion Paper No. 03/12, University of Nottingham - School of Economics, United Kingdom (2003)
12. Fershtman, C.: The interdependence between ownership status and market structure: the case of privatization. Economica **57**(22), 319–328 (1990)
13. Matsumura, T.: Stackelberg mixed duopoly with a foreign competitor. Bull. Econ. Res. **55**, 275–287 (2003)
14. Matsushima, N., Matsumura, T.: Mixed oligopoly and spatial agglomeration. Can. J. Econ. **36**(1), 62–87 (2004)
15. Matsumura, T., Kanda, O.: Mixed oligopoly at free entry markets. J. Econ. **84**(1), 27–48 (2005)
16. Ireland, N.J., Law, P.J.: The Economics of Labour-Managed Enterprises. Croom Helm, London (1982)
17. Bonin, J.P., Putterman, L.G.: Economics of Cooperation and the Labor-managed Economy. Harwood Academic Publishers, Switzerland (1987)
18. Stephen, F.H. (ed.): The Performance of Labour-Managed Firms. Palgrave Macmillan, London (1982)
19. Putterman, L.: Labour-managed firms. In: Durlauf, S.N., Blume, L.E. (eds.) The New Palgrave Dictionary of Economics, vol. 4, pp. 791–795. Palgrave Macmillan, Hampshire (2008)
20. Saha, B., Sensarma, R.: State ownership, credit risk and bank competition: a mixed oligopoly approach. Macroecon. Financ. Emerg. Mark. Econ. **6**(1), 1–13 (2013)
21. Mumcu, A., Oğur, S., Zenginobuz, U.: Competition between regulated and non-regulated generators on electric power networks. In: MPRA Paper 376. University Library of Munich, Germany (2001)
22. Kalashnykova, N.I., Bulavsky, V.A., Kalashnikov, V.V., Castillo-Pérez, F.J.: Consistent conjectural variations equilibrium in a mixed duopoly. J. Adv. Comput. Intell. Intell. Inform. **15**(4), 425–432 (2011)
23. Kalashnikov, V.V., Kalashnykova, N.I., Camacho, J.F.: Partially mixed duopoly and oligopoly: consistent conjectural variations equilibrium (ccve). part 1. In: Leyva-López, J.C., et al. (eds.) Fourth International Workshop on Knowledge Discovery, Knowledge Management and Decision Support. Atlantis Press, Mexico , pp. 198–206 (2013)
24. Kalashnikov, V.V., Bulavsky, V.A., Kalashnykova, N.I., Watada, J., Hernández-Rodríguez, D.J.: Analysis of consistent equilibria in a mixed duopoly. J. Adv. Comput. Intell. Intell. Inform. **18**(6), 962–970 (2014)
25. Liu, Y.F., Ni, Y.X., Wu, F.F., Cai, B.: Existence and uniqueness of consistent conjectural variation equilibrium in electricity markets. Int. J. Electr. Power Energy Syst. **29**(6), 455–461 (2007)
26. Bulavsky, V.A., Kalashnikov, V.V.: One-parametric method to study equilibrium. Econ. Math. Methods (Ekonomika i Matematicheskie Metody) **30**, 129–138 (1994). In Russian
27. Bulavsky, V.A., Kalashnikov, V.V.: Equilibrium in generalized Cournot and Stackelberg models. Econ. Math. Methods (Ekonomika i Matematicheskie Metody) **31**, 164–176 (1995). In Russian
28. Isac, G., Bulavsky, V.A., Kalashnikov, V.V.: Complementarity, Equilibrium, Efficiency and Economics. Kluwer Academic Publishers, Dordrecht (2002)

29. Kalashnikov, V.V., Kemfert, C., Kalashnikov-Jr, V.V.: Conjectural variations equilibrium in a mixed duopoly. Eur. J. Oper. Res. **192**(3), 717–729 (2009)
30. Kalashnikov, V.V., Cordero, A.E., Kalashnikov-Jr, V.V.: Cournot and Stackelberg equilibrium in mixed duopoly models. Optimization **59**(5), 689–706 (2010)
31. Kreps, D.M., Scheinkman, J.A.: Quantity precommitment and Bertrand competition yield Cournot outcomes. Bell J. Econ. **14**(2), 326–337 (1983)
32. Murphy, F.H., Smeers, Y.: Generation capacity expansion in imperfectly competitive restructured electricity markets. Oper. Res. **53**(4), 646–661 (2005)
33. Kimbrough, S.O., Murphy, F.H., Smeers, Y.: Extending Cournot: when does insight dissipate? Paper No. 14-036, Fox School of Business Research (2014)
34. Allaz, B., Vila, J.L.: Cournot competition, forward markets and efficiency. J. Econ. Theory **59**(1), 1–16 (1993)
35. Driskill, R.A., McCafferty, S.: Dynamic duopoly with output adjustment costs in international markets: taking the conjecture out of conjectural variations. In: Feenstra, R.C. (ed.) Trade Policies for International Competitiveness, pp. 125–144. University of Chicago Press, United States (1989)
36. Kalashnikov, V.V., Bulavsky, V.A., Kalashnykova, N.I., López-Ramos, F.: Consistent conjectures are optimal Cournot-Nash strategies in the meta-game. Optimization **66**(12), 2007–2024 (2017)
37. Kalashnykova, N.I., Bulavsky, V.A., Kalashnikov, V.V.: Consistent conjectures as optimal nash strategies in the upper level game. ICIC Express Lett. **6**(4), 965–970 (2012)
38. Kress, D., Pesch, E.: Competitive location and pricing on networks with random utilities. Netw. Spat. Econ. **16**(3), 837–863 (2016)
39. Braess, D.: Über ein Paradoxon aus der Verkehrsplanung. Unternehmensforschung **12**, 258–268 (1969)
40. Braess, D., Nagurney, A., Wakolbinger, T.: On a paradox of traffic planning. INFORMS **39**(4), 446–450 (2005)
41. Magnanti, T.L., Wong, R.T.: Network design and transportation planning: models and algorithms. Transp. Sci. **18**(1), 1–55 (1984)
42. Marcotte, P.: Network design problem with congestion effects: a case of bilevel programming. Math. Program. **34**(2), 142–162 (1986)
43. Dempe, S., Lohse, S.: Best highway toll assigning models and an optimality test (in German), Preprint, TU Bergakademie Freiberg, Nr. 2005-6, Fakultät für Mathematik und Informatik, Freiberg (2005)
44. Didi-Biha, M., Marcotte, P., Savard, G.: Path-based formulation of a bilevel toll setting problem. In: Dempe, S., Kalashnikov, V.V. (eds.) Optimization with Multi-Valued Mappings: Theory, Applications and Algorithms, pp. 29–50. Springer Science, Boston, MA (2006)
45. Dempe, S., Starostina, T.: Optimal toll charges: fuzzy optimization approach. In: Heyde, F., Löhne, A., Tammer, C. (eds.) Methods of Multicriteria Decision - Theory and Applications, pp. 29–45. Shaker Verlag, Aachen (2009)
46. Labbé, M., Marcotte, P., Savard, G.: On a class of bilevel programs. In: Di Pillo, G., Giannessi, F. (eds.) Nonlinear Optimization and Related Topics, pp. 183–206. Kluwer Academic Publishers, Dordrecht (2000)
47. Brotcorne, L.: Operational and strategic approaches to traffic routers' problems (in French), Ph.D. thesis, Université Libre de Bruxelles (1998)
48. Kalashnikov, V.V., Kalashnykova, N.I., Herrera, R.C.: Solving bilevel toll optimization problems by a direct algorithm using sensitivity analysis. In: Proceedings of the 2011 New Orleans International Academic Conference. New Orleans, LA, pp. 1009–1018 (2011)
49. Labbé, M., Marcotte, P., Savard, G.: A bilevel model of taxation and its applications to optimal highway pricing. Manag. Sci. **44**(12), 1608–1622 (1998)
50. Roch, S., Savard, G., Marcotte, P.: Design and analysis of an algorithm for stackelberg network pricing. Networks **46**(1), 57–67 (2005)
51. Brotcorne, L., Cirinei, F., Marcotte, P., Savard, G.: An exact algorithm for the network pricing problem. Discrete Optim. **8**(2), 246–258 (2011)

52. Brotcorne, L., Cirinei, F., Marcotte, P., Savard, G.: A tabu search algorithm for the network pricing problem. Comput. Oper. Res. **39**(11), 2603–2611 (2012)
53. Dempe, S., Zemkoho, A.B.: Bilevel road pricing; theoretical analysis and optimality conditions. Ann. Oper. Res. **196**(1), 223–240 (2012)
54. Kalashnikov, V.V., Herrera, R.C., Camacho, F., Kalashnykova, N.I.: A heuristic algorithm solving bilevel toll optimization problems. Int. J. Logist. Manag. **27**(1), 31–51 (2016)
55. Wan, Z., Yuan, L., Chen, J.: A filled function method for nonlinear systems of equalities and inequalities. Comput. Appl. Math. **31**(2), 391–405 (2012)
56. Wu, Z.Y., Mammadov, M., Bai, F.S., Yang, Y.J.: A filled function method for nonlinear equations. Appl. Math. Comput. **189**(2), 1196–1204 (2007)
57. Kalashnikov, V.V., Camacho, F., Askin, R., Kalashnykova, N.I.: Comparison of algorithms solving a bilevel toll setting problem. Int. J. Innov. Comput. Inf. Control **6**(8), 3529–3549 (2010)

Chapter 2
Consistent Conjectural Variations Equilibrium in a Semi-mixed Duopoly

2.1 Model Specification

Consider a semi-mixed duopoly with two producers where $i = 0$ is a semi-public company and $i = 1$ is a private firm. The companies supply a homogeneous produce under the expenditure estimated by the *cost functions* $f_i(q_i), i = 0, 1$, where $q_i \geq 0$ is the *output volume* by producer i. The market-clearing supply is specified by a demand (inverse price) function $G = G(p)$, whose argument p is the price suggested by the suppliers. An active demand D is nonnegative and independent of the price. The equilibrium between the demand and supply for a given price p is described by the following *balance equality*:

$$q_0 + q_1 = G(p) + D. \tag{2.1}$$

We assume the following properties of the model's data.

A1 The demand function $G = G(p) \geq 0$ has finite values for all $p \geq 0$, and is continuously differentiable with $G'(p) < 0$.

A2 For each $i = 0, 1$, the cost function $f_i(q_i)$ is quadratic with zero overhead costs, i.e.,

$$f_i(q_i) = \frac{1}{2}a_i q_i^2 + b_i q_i, \tag{2.2}$$

where

$$a_i > 0,\ b_i > 0, \quad i = 0, 1. \tag{2.3}$$

Moreover, it is assumed that

$$b_0 \leq b_1. \tag{2.4}$$

J. G. Flores Muñiz et al., *Public Interest and Private Enterprize: New Developments*, Lecture Notes in Networks and Systems 138,
https://doi.org/10.1007/978-3-030-58349-1_2

The private firm $i = 1$ selects his output $q_1 \geq 0$ in order to maximize its net profit function

$$\pi_1(p, q_1) = pq_1 - f_1(q_1), \tag{2.5}$$

whereas the semi-public company $i = 0$ decides its output volume $q_0 \geq 0$ so as to maximize the convex combination of *domestic social surplus* and the net profit function

$$S(p, q_0, q_1) = \beta \left(\int_0^{q_0+q_1} p(x)dx - pq_1 - f_0(q_0) \right) + (1-\beta)(pq_0 - f_0(q_0)), \tag{2.6}$$

where $0 < \beta \leq 1$.

Here, domestic social surplus involving the integral in (2.6) is usually interpreted as the money gained by the (domestic) consumer when he/she acquires the good at the lower price (established in the market) than that expected by him/her *before* the semi-public company entered the market (see the more detailed interpretation by the well-known Japanese mathematicians and economists: [1–3]).

According to our concept of conjectural variations equilibrium (CVE), we assume that both producers (semi-public and private) *conjecture* about variations in the market-clearing price p as a function of the perturbations in their output quantities. In terms of the first derivative, the latter assumption might be described by a *conjectured* dependency of (infinitesimal) affine variations of the price p upon (infinitely small) perturbations of the supply quantities q_i. Within this framework, the first order optimum condition depicting equilibrium reduces to the form: for the semi-public company ($i = 0$)

$$\frac{\partial S}{\partial q_0} = p - [\beta q_1 - (1-\beta)q_0]\frac{\partial p}{\partial q_0} - f_0'(q_0) \begin{cases} = 0, & \text{if } q_0 > 0, \\ \leq 0, & \text{if } q_0 = 0. \end{cases} \tag{2.7}$$

A similar first order optimality condition for the private producer ($i = 1$) yields

$$\frac{\partial \pi_1}{\partial q_1} = p + q_1\frac{\partial p}{\partial q_1} - f_1'(q_1) \begin{cases} = 0, & \text{if } q_1 > 0, \\ \leq 0, & \text{if } q_1 = 0. \end{cases} \tag{2.8}$$

On that account, in order to predict the (instantaneous) behavior of supplier i, one need make use of the first order derivative

$$\frac{\partial p}{\partial q_i} \equiv -v_i \tag{2.9}$$

rather than the (exact) functional dependency of p upon q_i. Even more, the latter dependency is extremely hard to estimate in a many-person game with several decision-makers. Here, the negative sign is applied in order to have nonnegative values of ν_i. Surely, the conjectured (first-order) dependency of p upon q_i should guarantee the concavity of the i-th producer's conjectured profit as a function of its supply, which implies the maximum of the producer's revenue. Under the assumption that the cost functions $f_i(q_i)$ are quadratic and strictly convex (assumption **A**1), it suffices supposing the coefficient ν_i (referred to as the i-th producer's *influence coefficient*) to be nonnegative and constant. In this case, the conjectured dependency of the semi-public company's objective function variations upon the production output $q_0 + \eta_0$ has the form

$$\begin{aligned}\hat{S}(q_0+\eta_0) =& \beta\left[\int\limits_0^{q_0+q_1+\eta_0} p(x)dx - (p - \nu_0\eta_0)q_1 - f_0(q_0+\eta_0)\right] \\ &+ (1-\beta)[(p-\nu_0\eta_0)(q_0+\eta_0) - f_0(q_0+\eta_0)],\end{aligned} \tag{2.10}$$

while the local maximum condition at $\eta_0 = 0$ is provided by the relation:

$$\begin{cases} p = -\nu_0\left[\beta q_1 - (1-\beta)q_0\right] + a_0 q_0 + b_0, & \text{if } q_0 > 0, \\ p \le -\beta\nu_0 q_1 + b_0, & \text{if } q_0 = 0. \end{cases} \tag{2.11}$$

Similarly, the private firm's profit presumes a local dependency upon the production output $q_1 + \eta_1$ in the form

$$\hat{\pi}_1(q_1+\eta_1) = (p - \nu_1\eta_1)(q_1+\eta_1) - f_1(q_1+\eta_1), \tag{2.12}$$

which permits formulating the maximum condition at $\eta_1 = 0$ as follows:

$$\begin{cases} p = \nu_1 q_1 + a_1 q_1 + b_1, & \text{if } q_1 > 0, \\ p \le b_1, & \text{if } q_1 = 0. \end{cases} \tag{2.13}$$

If the producers' conjectures concerning market-clearing price were assigned exogenously (like it was done in [4, 5]), one might assume the values ν_i to be functions of q_i and p. Notwithstanding that, we use the approach from [6], where the conjectures in the equilibrium are determined simultaneously with the market-clearing price p and the output volumes q_i by a special verification procedure. In this circumstance, the influence coefficients are the solution of a nonlinear system of equations found for the equilibrium only. In Sect. 2.3, such equilibrium state is referred to as *interior* and is described by the extended vector $(p, q_0, q_1, \nu_0, \nu_1)$.

2.2 Exterior Equilibrium

In order to describe the verification procedure, we need first to define a more elementary concept of equilibrium referred to as *exterior* (cf., [6]) with the parameters v_i assigned exogenously.

Definition 2.1 A vector (p, q_0, q_1) is called *exterior equilibrium* for the fixed coefficients $v_i \geq 0$, $i = 0, 1$, if the market is balanced, i.e., Eq. (2.1) is satisfied, and the optimality conditions for the semi-public company and private firm, (2.11) and (2.13) respectively, hold.

In what follows, we are going to consider only the case when the set of producers with strictly positive output volumes is fixed, i.e., it does not depend upon the values v_i of the influence coefficients. In order to assure this feature, we make the following assumption:

A3 For the price $p_0 = b_1$, the following inequality holds:

$$\frac{p_0 - b_0}{a_0} < G(p_0). \tag{2.14}$$

The latter assumption, together with **A**1 and **A**2 guarantees that, for all nonnegative values of v_i, $i = 0, 1$, there always exists a unique solution of the optimality conditions (2.11) and (2.13) satisfying the balance equation (2.1), i.e, the exterior equilibrium. Moreover, the latter three conditions (2.1), (2.11) and (2.13) can hold simultaneously if and only if $p > p_0$, or equivalently, if and only if $q_i > 0$ for all $i = 0, 1$.

Lemma 2.1 *Let assumptions A1–A3 be valid. Then, for all nonnegative values of* v_i, $i = 0, 1$, *supply values* q_i *are strictly positive (i.e.,* $q_i > 0$, $i = 0, 1$*) at any exterior equilibrium if and only if* $p > p_0$.

Comment. For reader's convenience, all the proofs are placed in the special Appendix.
We are now in a position to formulate the main result of this section.

Theorem 2.1 *Under assumptions A1–A3, for any* $\beta \in (0, 1]$, $D \geq 0$ *and* $v_i \geq 0$, $i = 0, 1$, *there exists uniquely the exterior equilibrium* (p, q_0, q_1) *depending continuously upon the parameters* (D, v_0, v_1). *The equilibrium price* p *as a function of these parameters is continuously differentiable with respect to* D *and* v_i, $i = 0, 1$. *Moreover* $p(D, v_0, v_1) > p_0$ *and*

$$\frac{\partial p}{\partial D} = \frac{1}{\dfrac{1}{(1-\beta)v_0 + a_0} + \dfrac{v_0 + a_0}{(1-\beta)v_0 + a_0}\left(\dfrac{1}{v_1 + a_1}\right) - G'(p)}. \tag{2.15}$$

2.3 Interior Equilibrium

Now we are in a position to define the concept of interior equilibrium. To do that, we first describe the procedure of verification of the influence coefficients ν_i exactly as it was introduced in [6]. Assume that the system is in the exterior equilibrium (p, q_0, q_1) that occurs for some given β, D and $\nu_i, i = 0, 1$. One of the producers, say $k, k \in \{0, 1\}$, temporarily changes its behavior by abstaining from the maximization of its conjectured objective function, şubtracts its produce q_k from the total demand and makes infinitesimal fluctuations around the latter. In mathematical terms, this is equivalent to restricting the list of producers to the subset $I_{-k} = \{1 - k\}$ with the output q_k subtracted from the active demand and the balance equation restated in the form:

$$q_{1-k} = G(p) + D - q_k. \tag{2.16}$$

Then, variations of the production output by producer k are equivalent to the corresponding active demand fluctuations in the form

$$dD_k = d(D - q_k) = -dq_k. \tag{2.17}$$

If we treat these variations as infinitesimal, we can assume that by observing the corresponding variations of the equilibrium price in the equilibrium attained among the remaining participants, producer k can evaluate the derivative of the equilibrium price with respect to the active demand, i.e, its influence coefficient.

Applying formula (2.15) from Theorem 2.1 to calculate the derivatives, one has to remember that producer k is temporally absent from the equilibrium model, hence, one has to exclude the terms corresponding to $i = k$ from the denominator. Having that in mind, we obtain the following criterion:

Definition 2.2 (*Consistency Criterion*) In the exterior equilibrium (p, q_0, q_1), the influence coefficients ν_i, $i = 0, 1$, are named *consistent* if the following equalities hold:

$$\nu_0 = \frac{1}{\dfrac{1}{\nu_1 + a_1} - G'(p)} \tag{2.18}$$

and

$$\nu_1 = \frac{1}{\dfrac{1}{(1-\beta)\nu_0 + a_0} - G'(p)}. \tag{2.19}$$

We are now ready to define the concept of interior equilibrium.

Definition 2.3 A vector $(p, q_0, q_1, \nu_0, \nu_1)$ is called *interior equilibrium* if, for the influence coefficients $\nu_i \geq 0$, $i = 0, 1$, the vector (p, q_0, q_1) is the exterior equilibrium, and the Consistency Criterion is valid for all ν_i, $i = 0, 1$.

The conjectural variations equilibrium is called *consistent* if the corresponding influence coefficients meet the Consistency Criterion (Definition 2.2).

Theorem 2.2 *Under assumptions A1–A3, there exists the interior equilibrium.*

Now, express the demand function's derivative with $\tau = G'(p)$, and replace the consistency conditions (2.18) and (2.19) with the formulas below:

$$\nu_0 = \frac{1}{\dfrac{1}{\nu_1 + a_1} - \tau} \tag{2.20}$$

and

$$\nu_1 = \frac{1}{\dfrac{1}{(1-\beta)\nu_0 + a_0} - \tau}, \tag{2.21}$$

where $\tau \in (-\infty, 0]$.

When $\tau \to -\infty$ the solutions of system (2.20) and (2.21) tend the (unique) limit solution $\nu_i = 0, i = 0, 1$. For any finite values of τ, we establish the following proposition.

Proposition 2.1 *For all $\tau \leq 0$, there exists a unique solution $\nu_i = \nu_i(\tau), i = 0, 1$, of system* (2.20) *and* (2.21), *which continuously depends upon τ. In addition, $\nu_i(\tau) \to 0$ whenever $\tau \to -\infty$, and $\nu_i(\tau)$ strictly grows and tends to $\nu_i(0)$ as $\tau \to 0, i = 0, 1$.*

2.4 A Special Case: An Affine Demand Function

Assume that the demand function $G(p)$ is affine, i.e.,

$$G(p) := -Kp + T, \tag{2.22}$$

where, $K > 0, T > 0$.

Under this extra assumption, Theorem 2.2 entails the next corollary.

Corollary 2.1 *Under assumptions A1–A3, for all $\beta \in (0, 1]$, the demand function of type* (2.22) *implies the uniqueness of the interior equilibrium.*

This section mainly targets at the study of the behavior (as a function of the parameter β) of the three most popular equilibrium kinds: (1) the consistent conjectural variations equilibrium (CCVE), (2) the Cournot-Nash equilibrium, and (3) the perfect competition equilibrium.

2.4.1 Consistent Conjectural Variations Equilibrium

The conjectural variation equilibrium (CVE) is called consistent if the influence coefficients at the interior CVE meet the consistency principle represented by systems (2.18) and (2.19).

For all $\beta \in (0, 1]$, Corollary 2.1 provides for the existence of the unique interior equilibrium $(p^*(\beta), q_0^*(\beta), q_1^*(\beta), v_0^*(\beta), v_1^*(\beta))$. The behavior of the CCVE is described in the following result:

Theorem 2.3 *For the affine demand function $G(p)$ from (2.22), the price $p^*(\beta)$, the supply outputs $q_i^*(\beta)$, $i = 0, 1$, and the influence coefficients $v_i^*(\beta)$, $i = 0, 1$, characterizing the interior equilibrium, together with total market supply $G^*(\beta) = q_0^*(\beta) + q_1^*(\beta)$, are continuously differentiable by $\beta \in (0, 1]$. Furthermore, $q_0^*(\beta)$ and $G^*(\beta)$ strictly increase, whereas $p^*(\beta)$, $v_0^*(\beta)$, $v_1^*(\beta)$ and $q_1^*(\beta)$ strictly decrease.*

2.4.2 Cournot-Nash Equilibrium

Below, we will examine the comportment of the (exterior) Cournot-Nash equilibrium as a function of the parameter β.

The well-known Cournot-Nash conjecture

$$\omega_i = \frac{\partial G}{\partial q_i} = 1, \quad i = 0, 1, \tag{2.23}$$

in the proposed framework is equivalent to the next conjecture:

$$v_i = -\frac{\partial p}{\partial q_i} = -\frac{1}{G'(p)} = -\frac{1}{K}, \quad i = 0, 1. \tag{2.24}$$

Theorem 2.1 implies that, for all $\beta \in (0, 1]$, there exists uniquely the Cournot-Nash equilibrium, denoted by $(p^c(\beta), q_0^c(\beta), q_1^c(\beta))$. The behavior of the Cournot-Nash equilibrium is described in the following result:

Theorem 2.4 *For the affine demand function $G(p)$ described in (2.22), the price $p^c(\beta)$ and the supply values $q_i^c(\beta)$, $i = 0, 1$, from the Cournot-Nash equilibrium, are continuously differentiable with respect to $\beta \in (0, 1]$. Moreover, $p^c(\beta)$ and $q_1^c(\beta)$ strictly decrease, whereas $q_0^c(\beta)$ strictly increase.*

It is quite evident that the Cournot-Nash equilibrium in our framework does not satisfy the Consistency Criterion, i.e., it is not interior (consistent) equilibrium.

2.4.3 Perfect Competition Equilibrium

Now, the comportment of the (exterior) perfect competition equilibrium as a function of the parameter β will be evaluated.

The perfect competition conjecture

$$\omega_i = 0, \quad i = 0, 1, \tag{2.25}$$

in our framework is described with the subsequent conjecture:

$$\nu_i = \frac{\partial p}{\partial q_i} = 0, \quad i = 0, 1. \tag{2.26}$$

For every $\beta \in (0, 1]$, Theorem 2.1 guarantees that there exist uniquely the exterior equilibrium implementing the perfect competition. The latter will be represented by $(p^t(\beta), q_0^t(\beta), q_1^t(\beta))$. Once more, in order to conduct a comparative study of all three types of equilibrium in our affine framework, we will establish the clear-cut formulas for the perfect equilibrium supplies and price in the next result.

Theorem 2.5 *For the affine demand function $G(p)$ described in* (2.22)*, the price $p^t(\beta)$ and the output volumes $q_i^t(\beta)$, $i = 0, 1$, related to the perfect competition equilibrium, are invariant for all $\beta \in (0, 1]$ and are described by the clear-cut expressions:*

$$p^t = \frac{a_0 b_1 + a_1 b_0 + a_0 a_1 (T + D)}{a_0 + a_1 + a_0 a_1 K}, \tag{2.27}$$

$$q_0^t = \frac{a_1 (G(b_0) + D) + (b_1 - b_0)}{a_0 + a_1 + a_0 a_1 K}, \tag{2.28}$$

$$q_1^t = \frac{a_0 (G(b_1) + D) - (b_1 - b_0)}{a_0 + a_1 + a_0 a_1 K}. \tag{2.29}$$

Like in the Cournot-Nash case, one can easily see that the perfect competition equilibrium within our framework does not meet the Consistency Criterion, thus, it is non-interior (inconsistent) equilibrium.

2.4.4 Comparison of Consistent CVE with Cournot-Nash and Perfect Competition Equilibriums

In this subsection, taking advantage of the simple (affine) demand function, we will deduce some comparative statics results. These are always interesting for evaluating the strong and weak points of different concepts of more or less similar nature.

Theorem 2.6 *For the affine demand function $G(p)$ from (2.22), the price functions in the CCVE, $p^*(\beta)$, the Cournot-Nash equilibrium, $p^c(\beta)$, and the perfect competition equilibrium, p^t, satisfy the following inequalities:*

$$p^t < \lim_{\beta \to 0} p^*(\beta), \tag{2.30}$$

and

$$p^*(\beta) < p^c(\beta), \quad \forall \beta \in (0, 1]. \tag{2.31}$$

Inequality (2.30), in general, does not hold when $\beta \to 1$. The latter is a very curious result because the perfect competition equilibrium price p^t is usually the lowest in the market while the Cournot-Nash equilibrium price is the highest (as seen in (2.31)). Moreover, in some cases, it may happen that $p^t > p^c(\beta)$ for the values of β near 1 (e.g., see Table 2.4).

2.4.5 Optimality Criterion for β

In order to find an optimal (in some sense) value for the "socialization" level β of the semi-public company, we study the behavior of the private firm's profit function in the two equilibrium states: CCVE and Cournot-Nash equilibrium.

The function $\pi_1(p, q_1)$ given by (2.5) is continuously differentiable with respect to p and q_1, while $p^*(\beta)$, $q_1^*(\beta)$, $p^c(\beta)$ and $q_1^c(\beta)$, are continuously differentiable with respect to β. Therefore, for the equilibrium states CCVE and Cournot-Nash, we have that the private firm's net profit values,

$$\pi_1^*(\beta) = p^*(\beta)q_1^*(\beta) - \frac{1}{2}a_1 q_1^*(\beta)^2 - b_1 q_1^*(\beta), \tag{2.32}$$

in the interior equilibrium (CCVE), as well as the similar values

$$\pi_1^c(\beta) = p^c(\beta)q_1^c(\beta) - \frac{1}{2}a_1 q_1^c(\beta)^2 - b_1 q_1^c(\beta), \tag{2.33}$$

in the Cournot-Nash exterior equilibrium, are continuously differentiable by $\beta \in (0, 1]$.

Theorem 2.7 *The functions $\pi_1^*(\beta)$ and $\pi_1^c(\beta)$ are strictly decreasing with respect to $\beta \in (0, 1]$. Moreover, the following inequalities hold:*

$$\pi_1^*(1) > \pi_1^c(1) \tag{2.34}$$

and

$$\lim_{\beta \to 0} \pi_1^*(\beta) < \lim_{\beta \to 0} \pi_1^c(\beta). \tag{2.35}$$

Directly from the proof of the last theorem, we conclude that there exists the value $\overline{\beta}$ such that $\pi_1^*(\overline{\beta}) = \pi_1^c(\overline{\beta})$. We now assume that the semi-public firm is socially responsible, and making use of the subsidy policy, it economically motivates the private firm to change its Cournot-Nash strategy to the consistent CVE comportment, or pays subsidies to the consumers to compensate the highest price in the Cournot-Nash equilibrium. The choice of this parameter $\overline{\beta}$ allows the semi-public company not to pay subsidies either to the private company or to the consumers.

With this idea in mind, we introduce the following definition:

Definition 2.4 The value of the parameter $\overline{\beta} \in (0, 1)$ such that $\pi_1^*(\overline{\beta}) = \pi_1^c(\overline{\beta})$ is called *optimal socialization level*.

From Theorem 2.7, it follows immediately that, for the duopoly model considered in this paper, we can always find the optimal socialization level for the semi-public company. In other words, the following result has been established above:

Theorem 2.8 *Under assumptions A1–A3, there exists the value of* $\overline{\beta} \in (0, 1)$ *such that* $\pi_1^*(\overline{\beta}) = \pi_1^c(\overline{\beta})$. *In other words, the optimal socialization level exists.*

2.5 Numerical Results

In this section, we rely on the data of the numerical experiments exposed in the work of [7]. Here, we describe the experiments in more detail.

The inverse demand function is given by

$$p(G, D) = 50 - 0.02(G + D) = 50 - 0.02(q_0 + q_1). \tag{2.36}$$

Then, solving (2.36) for $G + D$ yields the demand function

$$G(p) + D = -50p + 2500. \tag{2.37}$$

The producers' cost functions are quadratic and are described by (2.2), where the values a_i and b_i are given in Table 2.1.

We calculate and compare three types of equilibrium: the consistent conjectural variations equilibrium (CCVE), the Cournot-Nash equilibrium, and the perfect competition equilibrium. The influence coefficients for the CCVE are determined by Eqs. (2.18) and (2.19). For the Cournot-Nash equilibrium, the influence coefficients

Table 2.1 Experiments' input data

i	b_i	a_i
0	2.0	0.02
1	1.75	0.0175
2	3.25	0.00834

are given by the equality (2.24), while for the perfect competition equilibrium, they have the value from (2.26).

Based on the data of Table 2.1, we proceed to perform the numerical experiments for the following three instances:

Experiment 1 *Firm* $i = 0$ *is public and firm* $i = 2$ *is private.*

Experiment 2 *Firm* $i = 0$ *is public and firm* $i = 1$ *is private.*

Experiment 3 *Firm* $i = 2$ *is public and firm* $i = 1$ *is private.*

In each instance, we handle the following notation for each of the three kinds of equilibrium: **CNE** for Cournot-Nash Equilibrium, **CCVE** for Consistent Conjectural Variations Equilibrium, and **PCE** for Perfect Competition Equilibrium.

2.5.1 Experiment 1

For this instance, firm $i = 0$ is semi-public and firm $i = 2$ is private, so the semi-public firm is stronger than the private firm; that is, the inequality $b_0 \leq b_1$ holds (assumption **A**2). The numerical results of this experiment are shown in Table 2.2.

From the results of Table 2.2, we see that the behavior of variables is as described in the theorems of the previous sections. The numerical results show that, for socialization levels $0 < \beta \leq 0.50$, the private firm's profit is higher in the Cournot-Nash equilibrium than in the CCVE, but for $0.75 \leq \beta \leq 1$, its profit is higher in the CCVE equilibrium than in the Cournot-Nash. Then, the optimal socialization level lies within the interval $0.50 < \beta_{\text{optimal}} < 0.75$. Furthermore, as a result of the numerical experiment, the approximate optimal value $\beta_{\text{optimal}} = 0.55262$ is found, for which the private firm's net profit is almost the same both in the Cournot-Nash (CNE) and the Consistent Conjectural Variations Equilibrium (CCVE). The corresponding CCVE (interior equilibrium) is presented as follows: $(p^*, q_0^*, q_1^*, \nu_0^*, \nu_1^*) = (18.07, 829.5, 766.8, 0.009830, 0.01099)$. This means that, if the semi-public company $i = 0$ accepts its objective function as a mixture of 55% of domestic social surplus and 45% of its (would-be) net profit, then, the private (foreign) competitor is indifferent to the choice of the Cournot-Nash or CCVE model to generate its supply because its net profit is the same in both cases. This can be considered as a win-win outcome for the local authorities since they need not either subsidize the consumers (in order to reimburse the higher price of the commodity in the Cournot-Nash equilibrium) nor pay compensation to the private firm for having accepted the CCVE equilibrium model (which uses to decrease the private firm's net profit as compared to that in the Cournot-Nash equilibrium).

In the previous sections, we made use of assumption **A**2. In the following two experiments, we will consider the case when this assumption is not met to see how our model behaves.

Table 2.2 Results of Experiment 1

		$\omega_i = -G'(p)\nu_i$		
		CNE	**CCVE**	**PCE**
$\beta = 0.25$	ω_0	1.0	0.49870	0.0
	ω_2	1.0	0.57870	0.0
	p	22.609318	19.486879	13.595420
	q_0	686.424683	710.318787	579.770996
	q_2	683.109314	815.337402	1240.458008
	G	1369.533936	1525.656250	1820.229004
	π_0	14124.014648	13194.797852	11644.427734
	π_2	11278.648438	10466.422852	6416.529297
$\beta = 0.50$	ω_0	1.0	0.49275	0.0
	ω_2	1.0	0.55480	0.0
	p	20.858637	18.324474	13.595420
	q_0	835.733032	808.190979	579.770996
	q_2	621.335083	775.585388	1240.458008
	G	1457.068115	1583.776367	1820.229004
	π_0	19391.527344	19203.304688	19927.513672
	π_2	9331.004883	9183.150391	6416.529297
$\beta = 0.75$	ω_0	1.0	0.48590	0.0
	ω_2	1.0	0.52850	0.0
	p	18.868885	17.108099	13.595420
	q_0	1005.430603	911.731628	579.770996
	q_2	551.125122	732.863464	1240.458008
	G	1556.555664	1644.595093	1820.229004
	π_0	25023.082031	25747.185547	28210.599609
	π_2	7341.368652	7916.434082	6416.529297
$\beta = 1.0$	ω_0	1.0	0.47835	0.0
	ω_2	1.0	0.50000	0.0
	p	16.587503	15.846337	13.595420
	q_0	1200.000122	1020.859924	579.770996
	q_2	470.624695	686.823181	1240.458008
	G	1670.624756	1707.683105	1820.229004
	π_0	31014.878906	32875.441406	36493.683594
	π_2	5353.354980	6684.358887	6416.529297

2.5.2 *Experiment 2*

In this instance, firm $i = 0$ is semi-public and firm $i = 1$ is private, so the semi-public firm is weaker than the private firm. The numerical results of this experiment are shown in Table 2.3.

Table 2.3 Results of Experiment 2

		$\omega_i = -G'(p)\nu_i$		
		CNE	**CCVE**	**PCE**
$\beta = 0.25$	ω_0	1.0	0.59445	0.0
	ω_2	1.0	0.59110	0.0
	p	23.938864	21.564388	17.181818
	q_0	711.353699	746.033264	759.090942
	q_2	591.703064	675.747375	881.818176
	G	1303.056763	1421.780640	1640.909180
	π_0	14790.943359	14083.678711	12493.647461
	π_2	10065.734375	9393.968750	6804.028809
$\beta = 0.50$	ω_0	1.0	0.59000	0.0
	ω_2	1.0	0.56425	0.0
	p	22.112148	20.202539	17.181818
	q_0	851.401917	848.829346	759.090942
	q_2	542.990601	641.043823	881.818176
	G	1394.392578	1489.873169	1640.909180
	π_0	19596.324219	19344.347656	19225.105469
	π_2	8476.324219	8233.186523	6804.028809
$\beta = 0.75$	ω_0	1.0	0.58485	0.0
	ω_2	1.0	0.53400	0.0
	p	20.010050	18.733961	17.181818
	q_0	1012.562927	960.608276	759.090942
	q_2	486.934662	602.693542	881.818176
	G	1499.497559	1563.301758	1640.909180
	π_0	24847.169922	25176.449219	25956.564453
	π_2	6816.779297	7057.777832	6804.028809
$\beta = 1.0$	ω_0	1.0	0.57895	0.0
	ω_2	1.0	0.50000	0.0
	p	17.565216	17.155014	17.181818
	q_0	1200.000122	1082.066895	759.090942
	q_2	421.739105	560.182312	881.818176
	G	1621.739258	1642.249268	1640.909180
	π_0	30578.642578	31659.882813	32688.021484
	π_2	5113.585938	5883.829590	6804.028809

The results shown in Table 2.3 demonstrate that the variables still behave as described in the theorems of the previous sections, even though assumption **A2** is not met.

In this second experiment, we have that $\beta_{\text{optimal}} = 0.62905$ and its corresponding interior equilibrium is as follows: $(p^*, q_0^*, q_1^*, \nu_0^*, \nu_1^*) = (19.46, 905.4, 627.2, 0.01175, 0.01098)$.

2.5.3 Experiment 3

For this instance, firm $i = 2$ is semi-public and firm $i = 1$ is private, so the semi-public firm now is even weaker than the private firm in comparison to the previous experiment. Again, we observe that the variables behave according to our model. The numerical results of this experiment are presented in Table 2.4.

Table 2.4 Results of Experiment 3

		$\omega_i = -G'(p)\nu_i$		
		CNE	**CCVE**	**PCE**
$\beta = 0.25$	ω_0	1	0.57135	0
	ω_2	1	0.4581	0
	p	21.52866	18.10005	13.16771
	q_0	896.135986	981.770264	1189.17383
	q_2	527.430908	613.227173	652.440613
	G	1423.5669	1594.99744	1841.6145
	π_0	18097.7637	16920.0293	14375.8008
	π_2	7997.77197	6735.87891	3724.68872
$\beta = 0.50$	ω_0	1	0.5625	0
	ω_2	1	0.41105	0
	p	18.86553	16.055681	13.16771
	q_0	1100.30933	1141.02332	1189.17383
	q_2	456.414124	556.192444	652.440613
	G	1556.72339	1697.21582	1841.6145
	π_0	24250.3164	23585.2266	22854.6621
	π_2	5989.02344	5249.89893	3724.68872
$\beta = 0.75$	ω_0	1	0.5519	0
	ω_2	1	0.3569	0
	p	15.652787	13.826693	13.16771
	q_0	1346.61963	1318.49219	1189.17383
	q_2	370.740997	490.173065	652.440613
	G	1717.3606	1808.66528	1841.6145
	π_0	31259.9844	31230.5996	31333.5254
	π_2	3951.65552	3817.31006	3724.68872
$\beta = 1.0$	ω_0	1	0.539	0
	ω_2	1	0.2943	0
	p	11.700713	11.425176	13.16771
	q_0	1649.61206	1515.0188	1189.17383
	q_2	265.352356	413.722351	652.440613
	G	1914.96436	1928.74121	1841.6145
	π_0	39263.8008	40014.6523	39812.3867
	π_2	2024.34131	2505.13232	3724.68872

For the third instance, we have that $\beta_{\text{optimal}} = 0.80324$ and its corresponding interior equilibrium is as follows: $(p^*, q_0^*, q_1^*, \nu_0^*, \nu_1^*) = (13.33, 1359, 474.8, 0.01099, 0.006886)$.

From the results of the numerical experiments, we can see that the weaker semi-public company (as compared to the private company), the closer to 1 its optimal socialization level.

References

1. Matsumura, T.: Stackelberg mixed duopoly with a foreign competitor. Bull. Econ. Res. **55**, 275–287 (2003)
2. Matsushima, N., Matsumura, T.: Mixed oligopoly and spatial agglomeration. Can. J. Econ. **36**(1), 62–87 (2004)
3. Matsumura, T., Kanda, O.: Mixed oligopoly at free entry markets. J. Econ. **84**(1), 27–48 (2005)
4. Bulavsky, V.A., Kalashnikov, V.V.: One-parametric method to study equilibrium. Econ. Math. Methods (Ekonomika i Matematicheskie Metody) **30**, 129–138 (1994). In Russian
5. Bulavsky, V.A., Kalashnikov, V.V.: Equilibrium in generalized Cournot and Stackelberg models. Econ. Math. Methods (Ekonomika i Matematicheskie Metody) **31**, 164–176 (1995). In Russian
6. Bulavsky, V.A.: Structure of demand and equilibrium in a model of oligopoly. Econ. Math. Methods (Ekonomika i Matematicheskie Metody) **33**, 112–134 (1997). In Russian
7. Liu, Y.F., Ni, Y.X., Wu, F.F., Cai, B.: Existence and uniqueness of consistent conjectural variation equilibrium in electricity markets. Int. J. Electr. Power Energy Syst. **29**(6), 455–461 (2007)

Chapter 3
Consistent Conjectural Variations Coincide with the Nash Solution in the Meta-Model

3.1 Model Specification

Consider an oligopoly of at least two producers of a homogeneous good with cost functions $f_i = f_i(q_i)$, $i = 1, \ldots, n$, $n \geq 2$, where $q_i \geq 0$ is the supply by producer i. Consumers' demand is described by a demand function $G = G(p)$, whose argument p is the market price established by a cleared market. An active demand D is nonnegative and does not depend upon the price. The equilibrium between supply and demand for a given price p is guaranteed by the following balance equality:

$$\sum_{i=1}^{n} q_i = G(p) + D. \tag{3.1}$$

We assume the following properties of the model's data:

A4 The demand function $G = G(p) \geq 0$ is defined for $p > 0$, being strictly decreasing and continuously differentiable.

A5 For each $i = 1, \ldots, n$, the function $f_i = f_i(q_i)$ is defined for every $q_i \geq 0$, is twice continuously differentiable, and in addition, the following inequalities hold:

$$f_i'(0) > 0 \text{ and } f_i''(q_i) > 0, \quad \forall q_i \geq 0. \tag{3.2}$$

Next, every producer $i = 1, \ldots, n$ chooses its output volume $q_i \geq 0$ so as to maximize its net profit function:

$$\pi_i(p, q_i) = pq_i - f_i(q_i). \tag{3.3}$$

J. G. Flores Muñiz et al., *Public Interest and Private Enterprize: New Developments*, Lecture Notes in Networks and Systems 138,
https://doi.org/10.1007/978-3-030-58349-1_3

Now we postulate that the producers admit that their perturbations in production volumes may affect the price value p. Thus, the first order maximum condition to describe the equilibrium will have the form:

$$\frac{\partial \pi_i}{\partial q_i} = p + q_i \frac{\partial p}{\partial q_i} - f_i'(q_i) \begin{cases} = 0, & \text{if } q_i > 0, \\ \leq 0, & \text{if } q_i = 0, \end{cases} \quad i = 1, \ldots, n. \tag{3.4}$$

Therefore, to describe the (infinitesimal) behavior of producer i, it is enough to conjecture the first order derivative

$$\frac{\partial p}{\partial q_i} \equiv -v_i. \tag{3.5}$$

The conjectured first-order dependence of p on q_i must provide, at least locally, concavity of the i-th producer's conjectured profit as a function of its output. As we suppose that the cost functions f_i are strictly convex and strictly increasing, by inequalities (3.2), then, for all $i = 1, \ldots, n$, concavity of the product pq_i with respect to q_i would suffice. Here, it is enough to assume the coefficient v_i to be nonnegative and constant. Then, the conjectured dependence of the profit's variations upon the production output $q_i + \eta_i$ has the form:

$$\widehat{\pi}_i(q_i + \eta_i) = (p - v_i \eta_i)(q_i + \eta_i) - f_i(q_i + \eta_i), \tag{3.6}$$

which is a concave function of η_i.

Here, it is worthwhile to mention that relation (3.6) does *not* mean that producer i exercises its market power. In fact, vice versa, player i is a price-taker: it accepts its *conjectured* influence coefficient v_i and calculates the expected variation in the market-clearing price p (and hence in its net profit) under the infinitesimal variation of its produce η_i.

Therefore, the maximum necessary condition at $\eta_i = 0$ is provided by the relationships

$$\begin{cases} p = v_i q_i + f_i'(q_i), & \text{if } q_i > 0, \\ p \leq f_i'(0), & \text{if } q_i = 0, \end{cases} \quad i = 1, \ldots, n, \tag{3.7}$$

and it is the sufficient condition, too.

Here we use, again, the approach from [1, 2], where the conjectured parameters for the equilibrium are determined simultaneously with the price p and the output values q_i in the *interior* equilibrium state described by the combined vector $(p, q_1, \ldots, q_n, v_1, \ldots, v_n)$. Nevertheless, in order to define the interior equilibrium, we first need to introduce the notion of *exterior* equilibrium (cf., [1]) with the parameters v_i assigned in the exogenous form.

3.2 Exterior Equilibrium

We define the concept of exterior equilibrium as follows:

Definition 3.1 A vector $(p, q_1, \ldots, q_n)$ is called *exterior equilibrium* for the given influence coefficients $\nu_i \geq 0$, $i = 1, \ldots, n$, if the market is balanced, i.e., equality (3.1) holds, and for each $i = 1, \ldots, n$, the maximum conditions (3.7) are valid.

From now onward, we are going to consider only the case when the set of really producing participants is fixed. To guarantee this feature, we make the assumption listed below.

A6 For $p_0 = \max\limits_{1\leq i\leq n}\{f_i'(0)\}$ and any $i = 1, \ldots, n$, there exists a unique (due to **A5**) supply volume $q_i^0 \geq 0$ such that

$$p_0 = f_i'(q_i^0), \text{ and in addition, } \sum_{i=1}^{n} q_i^0 < G(p_0). \tag{3.8}$$

Lemma 3.1 *Assumptions A4–A6 imply that for all nonnegative values of ν_i, $i = 1, \ldots, n$, any exterior equilibrium has its supply values q_i strictly positive if and only if $p > p_0$.*

Proof See [3]. ■

The existence and uniqueness of the exterior equilibrium for any set of (nonnegative) conjectures (influence coefficients) were established in [1]. However, in the latter paper, only differentiability of the equilibrium clearing price p with respect to the active demand D was proven, while in this work, we also need to show that the same equilibrium price function $p = p(D, \nu_1, \ldots, \nu_n)$ is differentiable by the influence coefficients, too. Therefore, the following theorem has been proved.

Theorem 3.1 *Under assumptions A4–A6, for any $D \geq 0$, $\nu_i \geq 0$, $i = 1, \ldots, n$, there exists uniquely the exterior equilibrium $(p, q_1, \ldots, q_n)$ that depends continuously on the parameters $D \geq 0$, $\nu_i \geq 0$, $i = 1, \ldots, n$. The equilibrium price $p = p(D, \nu_1, \ldots, \nu_n)$, as a function of these parameters, is differentiable with respect to both D and ν_i, $i = 1, \ldots, n$. Moreover, $p(D, \nu_1, \ldots, \nu_n) > p_0$, and*

$$\frac{\partial p}{\partial D} = \frac{1}{\sum\limits_{i=1}^{n} \dfrac{1}{\nu_i + f_i''(q_i)} - G'(p)}, \tag{3.9}$$

while

$$\frac{\partial p}{\partial \nu_i} = \frac{\dfrac{q_i}{\nu_i + f_i''(q_i)}}{\sum\limits_{k=1}^{n} \dfrac{1}{\nu_k + f_k''(q_k)} - G'(p)} > 0, \quad i = 1, \ldots, n. \tag{3.10}$$

Similarly, the equilibrium supply $q_i = q_i(D, \nu_1, \ldots, \nu_n)$, $i = 1, \ldots, n$, is differentiable with respect to the influence coefficients ν_k, $k = 1, \ldots, n$, with the partial derivatives having the forms:

$$\frac{\partial q_i}{\partial \nu_i} = -\frac{q_i}{\nu_i + f_i''(q_i)} \left[\frac{\sum\limits_{\substack{k=1\\k\neq i}}^{n} \frac{1}{\nu_k + f_k''(q_k)} - G'(p)}{\sum\limits_{k=1}^{n} \frac{1}{\nu_k + f_k''(q_k)} - G'(p)} \right] < 0, \quad i = 1, \ldots, n, \tag{3.11}$$

and

$$\frac{\partial q_i}{\partial \nu_j} = \frac{1}{\nu_i + f_i''(q_i)} \left[\frac{\frac{q_j}{\nu_j + f_j''(q_j)}}{\sum\limits_{k=1}^{n} \frac{1}{\nu_k + f_k''(q_k)} - G'(p)} \right] > 0, \quad i, j = 1, \ldots, n, \ j \neq i. \tag{3.12}$$

Proof See [3]. ■

3.3 Interior Equilibrium

Now we are in a position to define the concept of interior equilibrium. To do that, we make use of the verification procedure for the influence coefficients ν_i introduced in [1]. Assume that the system is in the exterior equilibrium $(p, q_1, \ldots, q_n)$ that occurs for some given D and ν_i, $i = 1, \ldots, n$. Now, producer k, $1 \leq k \leq n$, temporarily changes its behavior by abstaining from maximizing its conjectured profit, subtracts its production q_k from the total demand and makes infinitesimal fluctuations around the latter. This is tantamount to restricting the model's producers to the subset $I_{-k} := \{i \mid 1 \leq i \leq n,\ i \neq k\}$ with the output q_k subtracted from the active demand with the balance equality restated in the form:

$$\sum_{\substack{i=0\\i\neq k}}^{n} q_i = G(p) + D - q_k. \tag{3.13}$$

Variations of the production output by producer k are, then, equivalent to the active demand fluctuation in the form $dD_k := d(D - q_k) = -dq_k$. If we treat these variations as infinitesimal, producer k can evaluate the derivative of the equilibrium price with respect to the active demand, i.e., their influence coefficient.

Applying formula (3.9) from Theorem 3.1 to calculate the derivatives, producer k is temporarily absent from the equilibrium model. Hence one has to exclude the term with number $i = k$ from the sum. This yields the Consistency Criterion.

Definition 3.2 (*Consistency Criterion*) In the exterior equilibrium $(p, q_1, \ldots, q_n)$, the influence coefficients $\nu_i, i = 1, \ldots, n$, are called *consistent* if the following equalities hold:

$$\nu_i = \frac{1}{\sum\limits_{\substack{j=1\\ j\neq i}}^{n} \dfrac{1}{\nu_j + f_j''(q_j)} - G'(p)}, \quad i = 1, \ldots, n. \tag{3.14}$$

Now, we define the consistent (interior) equilibrium.

Definition 3.3 A collection $(p, q_1, \ldots, q_n, \nu_1, \ldots, \nu_n)$ is referred to as *interior equilibrium* if, for the influence coefficients $\nu_i \geq 0,\ i = 1, \ldots, n,$ the vector $(p, q_1, \ldots, q_n)$ is the exterior equilibrium, and the Consistency Criterion 3.2 is valid for all those ν_i.

Theorem 3.2 *Let the number of oligopoly producers be at least three, i.e., $n \geq 3$, then, under assumptions* **A4**–**A6**, *there exists an interior equilibrium. Moreover, if the number of producers is two, i.e., $n = 2$, in addition to assumptions* **A4**–**A6**, *suppose that there exists an $\varepsilon > 0$ such that $G'(p) \leq -\varepsilon$ for all $p > 0$, then, there exists interior equilibrium.*

Without additional assumptions or simplifications of the model, uniqueness of the interior equilibrium is not guaranteed.

In our future research, we are going to extend the obtained results to the case of not necessarily differentiable demand functions. However, some of the essential techniques can be developed now, in the differentiable case but under slightly stronger assumptions about the structure of the producers' cost functions. Namely, let us introduce the following assumption instead of **A5**. Moreover, this new assumption is used in the extension of the existence Theorem 3.2 for the case of duopoly, i.e., when $n = 2$.

A7 For every $i = 1, \ldots, n$, the cost function f_i is quadratic (and strictly convex) with $f_i(0) = 0$, $f_i'(0) > 0$ and $f_i'' > 0$, i.e.,

$$f_i(q_i) = \frac{1}{2} a_i q_i^{\,2} + b_i q_i, \tag{3.15}$$

where $a_i > 0$, $b_i > 0$, $i = 1, \ldots, n$.

Theorem 3.3 *Let $n = 2$ (duopoly), and assumptions* **A4**, **A6** *and* **A7** *hold true. If in addition there exists $\varepsilon > 0$ such that $G'(p) \leq -\varepsilon$ for all $p > 0$, then, there exits the unique interior equilibrium.*

Proof See [3]. ■

Now denote the value of the demand function's derivative by $\tau = G'(p)$ and rewrite the consistency equations (3.14) in the form:

$$\nu_i = \frac{1}{\sum\limits_{\substack{j=1\\ j\neq i}}^{n} \frac{1}{\nu_j + f_j''(q_j)} - \tau}, \quad i = 1, \dots, n, \tag{3.16}$$

where $\tau \in (-\infty, 0]$.

When $\tau \to -\infty$, the system (3.16) converges to the solution $\nu_i = 0, i = 1, \dots, n$. For all finite values of τ we establish the following proposition.

Theorem 3.4 *Let assumptions A4, A6 and A7 be valid. Then, for any* $\tau \in (-\infty, 0]$ *there exists a unique solution* $\nu_i = \nu_i(\tau), i = 1, \dots, n$, *of system* (3.16), *continuously depending upon* τ. *Furthermore,* $\nu_i \to 0$ *when* $\tau \to -\infty$, *and* $\nu_i(\tau)$ *strictly increases and tends to* $\nu_i(0)$ *as* $\tau \to 0$, $i = 1, \dots, n$.

Proof is easily deduced from that of Theorem 3 in [4]. ∎

3.4 Consistent Conjectures as Optimal Nash Strategies in the Meta-Game

This section establishes the three most important results of this chapter. Indeed, under certain rather mild conditions, we prove the equivalence of the consistent conjectural variations equilibrium (CCVE) in the oligopoly model to the classical Cournot-Nash equilibrium in, what we call the meta-game. The latter comprises the same producers of the oligopoly but with their conjectures (about the possible price variations) as their strategies.

These results seem to be very interesting in two aspects. First, they could be considered as a good justification of the CVE concept as being tightly related to the classical Nash equilibrium. Second, this equivalence occurring in the oligopoly can help one develop a concept similar to the CVE but in application to other kinds of economic and financial models that lack some attributes of the oligopoly and thus do not allow one to introduce the consistent CVE directly. In other words, one could define the consistent CVE in such a model via the Nash equilibrium in the corresponding meta-game.

To begin with, Theorem 3.1 allows us to define the following many-person game $\Gamma = (N, V, \Pi, D)$, which will be referred to as the *meta-game*. Here, $D \geq 0$ is the fixed value of the active demand, $N = \{1, \dots, n\}$ is the set of the same producers (the players) as in the model described above, $V = \mathbb{R}_+^n$ represents the set of possible strategies, i.e., the vectors of conjectures $\nu = (\nu_1, \dots, \nu_n)$ accepted by the producers. Finally, $\Pi = \Pi(\nu) = (\pi_1, \dots, \pi_n)$ is the collection of payoff values defined (uniquely, according to Theorem 3.1) by the strategy vector ν. Indeed, the payoff values $\pi_i = \pi_i(\nu)$, $i = 1, \ldots, n$, are defined by formula (3.3), were the equilibrium outputs $q_i \geq 0$, $i = 1, \ldots, n$, as well as the equilibrium price p, are the elements of the exterior equilibrium whose existence and uniqueness is guaranteed by Theorem 3.1 from Sect. 3.2.

Now, the main results of this chapter are as follows. As was mentioned in the introduction, the Cournot-Nash conjectures $\omega_i = 1$ are usually inconsistent (in sense of criterion 3.2) in our single commodity market model. In other words, the Cournot-Nash conjectures $\nu_i = -1/G'(p)$ in general *do not* satisfy the consistency system (3.14). However, in the meta-game introduced above, the consistent conjectures ν_i, $i = 1, \ldots, n$, determined by (3.14) provide the Nash equilibrium. This curious fact could be considered as an extra argument supporting the concept of interior equilibrium introduced in Sect. 3.3.

Theorem 3.5 *Suppose that assumptions **A4–A6** hold. Then, any Nash equilibrium in the meta-game $\Gamma = (N, V, \Pi, D)$ generates interior equilibrium in the original oligopoly.*

Proof See [3]. ■

Since the meta-game strategies set $V = \mathbb{R}^n_+$ is unbounded, the existence of at least one Nash equilibrium state in this game is by no means easy to check. The following three results (under some extra assumptions) guarantee that the existence of interior equilibrium in the original oligopoly implies the existence of Nash equilibrium in the meta-game.

Theorem 3.6 *Suppose that the stronger assumption **A7** is true, together with **A4** and **A6**, and suppose that the function G is concave. Then, the Consistency Criterion for the original oligopoly is a necessary and sufficient condition for the collection of influence conjectures $\nu = (\nu_1, \ldots, \nu_n)$ to produce Nash equilibrium in the meta-game.*

Theorem 3.7 *In addition to assumptions **A4**, **A6** and **A7**, if the demand function is affine, that is,*

$$G(p) := -Kp + T, \tag{3.17}$$

where $K > 0$ and $T > 0$, then, the Consistency Criterion for the original oligopoly is a necessary and sufficient condition for the collection of influence conjectures $\nu = (\nu_1, \ldots, \nu_n)$ to form the Nash equilibrium in the meta-game.

Since the concavity of the demand function may be a much too restrictive requirement, the next theorem relaxes it by replacing it with the Lipschitz continuity of the derivative $G'(p)$.

Theorem 3.8 *Suppose that apart from assumptions **A4**, **A6** and **A7**, the regular demand function's derivative is Lipschitz continuous. In more detail, for $n \geq 3$ assume that for any $p_1 > 0$ and $p_2 > 0$ the following inequality holds:*

$$|G'(p_1) - G'(p_2)| \leq \frac{1}{2s^2 G(p_0)} |p_1 - p_2|, \tag{3.18}$$

where $s = \max\{a_1, \ldots, a_n\}$, and the price p_0 is the one defined in the assumption A6. Next, if $n = 2$ (duopoly), we again suppose that there exists $\varepsilon > 0$ such that $G'(p) \leq -\varepsilon$ for all $p > 0$, and the Lipschitz continuity of the demand function is described in the form:

$$|G'(p_1) - G'(p_2)| \leq \frac{2}{\left(\frac{a_1 + a_2}{\varepsilon \min\{a_1, a_2\}} + 3\max\{a_1, a_2\}\right)^2 G(p_0)} |p_1 - p_2|, \ \forall p_1, p_2 > 0. \tag{3.19}$$

Then, the Consistency Criterion for the original oligopoly is a necessary and sufficient condition for the collection of influence conjectures $\nu = (\nu_1, \ldots, \nu_n)$ to be the Nash equilibrium in the meta-game.

3.5 Numerical Experiments

Now we illustrate our main results from Sect. 3.4. For the numerical experiments, we consider the inverse demand and costs function from an electricity market presented in [5].

The inverse demand function is given by:

$$p(G, D) = 50 - 0.02(G + D), \tag{3.20}$$

thus, the demand function has the form:

$$G(p) + D = -50p + 2500. \tag{3.21}$$

There are $n = 6$ firms with quadratic costs functions, i.e., $f_i(q_i) = \frac{1}{2} a_i {q_i}^2 + b_i q_i$, $i = 1, \ldots, n$, where the coefficients a_i and b_i are given in Table 3.1.

Table 3.1 Quadratic costs function's coefficients

i	a_i	b_i
1	0.02	2.0
2	0.0175	1.75
3	0.025	3.0
4	0.025	3.0
5	0.0625	1.0
6	0.00834	3.25

Table 3.2 Interior equilibrium for the electricity market

i	1	2	3	4	5	6
ω_i	0.19275	0.19635	0.18759	0.18759	0.17472	0.22391
q_i	353.40	405.12	258.44	258.44	142.90	560.18
π_i	1730.4	2080.6	1085.4	1085.4	709.48	2713.8
p	10.431					
G	1978.5					

In addition, as a kind of dual concept developed in our previous papers, we are going to consider the influence of the producers over the total output G (cf., [6]), i.e, the influence coefficients $\omega_i := \dfrac{\partial G}{\partial q_i}, i = 1, \ldots, n.$

In this sense, the conjectures $\omega_i = 1, \forall i = 1, \ldots, n$, correspond to the Cournot-Nash conjecture, while the zero-conjectures $\omega_i = 0, \forall i = 1, \ldots, n$, lead to the perfect competition model.

Moreover, by the chain rule, one can easily verify the relationship

$$\omega_i = -G'(p)\nu_i, \quad \forall i = 1, \ldots, n. \tag{3.22}$$

Experiment 4 *For the electricity market described above, the producers' influence coefficients ω_i, supplies q_i and profits π_i, $i = 1, \ldots, n$, along with the market's price p and demand G, of the interior equilibrium results are shown in Table 3.2.*

Next, we vary the influence coefficient of one of the producers and compute the corresponding exterior equilibrium to see how their profits change.

In Tables 3.3, 3.4, 3.5, 3.6, 3.7 and 3.8 we see that whenever one of the producers (unilaterally) increases or decreases its consistent influence coefficient, its profit drops. This, as proved in Sect. 3.4, is due to the fact that the consistent influence

Table 3.3 Net profits when producer 1 (unilaterally) changes its consistent influence coefficient

i	1	2	3	4	5	6
ω_i	0.19275 **0.39165**	0.19635	0.18759	0.18759	0.17472	0.22391
π_i	1730.4 **1703.2**	2163.4	1136.0	1136.0	735.45	2844.6

Table 3.4 Net profits when producer 2 (unilaterally) changes its consistent influence coefficient

i	1	2	3	4	5	6
ω_i	0.19275	0.19635 **0.12000**	0.18759	0.18759	0.17472	0.22391
π_i	1688.8	2080.6 **2072.0**	1055.9	1055.9	694.22	2637.3

Table 3.5 Net profits when producer 3 (unilaterally) changes its consistent influence coefficient

i	1	2	3	4	5	6
ω_i	0.19275	0.19635	0.18759 **0.43615**	0.18759	0.17472	0.22391
π_i	1783.6	2142.7	1085.4 **1066.3**	1123.3	728.97	2811.9

Table 3.6 Net profits when producer 4 (unilaterally) changes its consistent influence coefficient

i	1	2	3	4	5	6
ω_i	0.19275	0.19635	0.18759	0.18759 **0.061570**	0.17472	0.22391
π_i	1697.1	2041.7	1061.8	1085.4 **1077.8**	697.27	2652.6

Table 3.7 Net profits when producer 5 (unilaterally) changes its consistent influence coefficient

i	1	2	3	4	5	6
ω_i	0.19275	0.19635	0.18759	0.18759	0.17472 **0.14870**	0.22391
π_i	1728.9	2078.8	1084.3	1084.3	709.48 **709.44**	2711.0

Table 3.8 Net profits when producer 6 (unilaterally) changes its consistent influence coefficient

i	1	2	3	4	5	6
ω_i	0.19275	0.19635	0.18759	0.18759	0.17472	0.22391 **0.47305**
π_i	1968.4	2358.2	1255.6	1255.6	796.41	2713.8 **2578.1**

coefficients describing the interior equilibrium form the Nash equilibrium in the meta-game.

Experiment 5 *Here, we are going to find the interior equilibrium for an electricity market (as the base model) that does* not *meet the conditions of Theorems 3.6 and 3.8, and test if it serves as the Nash equilibrium in the meta game.*

Now, we consider the following demand function:

$$G(p) + D = 2400p^{-1.2} + 1600, \tag{3.23}$$

along with the cost functions described above.

Table 3.9 Interior equilibrium for the electricity market with the demand function from (3.23)

i	1	2	3	4	5	6
ω_i	0.086234	0.088004	0.083674	0.083674	0.077221	0.10112
q_i	317.77	363.63	231.17	231.17	132.16	479.30
π_i	1467.8	1769.1	903.22	903.22	616.76	2179.9
p	9.7968					
G	1755.2					

Table 3.10 Net profits when producers change unilaterally their consistent strategy with the demand function from (3.23)

i	1	2	3	4	5	6
ω_i	**0.060438**	**0.076765**	**0.035183**	**0.079306**	**0.074729**	**0.093522**
	0.086234	0.088004	0.083674	0.083674	0.077221	0.10112
	0.11277	**0.10719**	**0.095928**	**0.12058**	**0.079999**	**0.1033**
π_i	**1464.0**	**1768.1**	**897.04**	**903.18**	**616.76**	**2178.5**
	1467.8	1769.1	903.22	903.22	616.76	2179.9
	1464.3	**1766.4**	**902.88**	**900.35**	**616.76**	**2179.8**

The function (3.23) is *not* concave and its derivative is *not* Lipschitz, but we can still compute the corresponding interior equilibrium. The corresponding data is shown in Table 3.9.

Finally, we vary the consistent influence coefficient of one of the producers and compute the corresponding exterior equilibrium to see how their profits change. These results are shown in Table 3.10.

In Table 3.10 one can see that each time one of the producers changes (unilaterally) its consistent influence coefficient, its profit decreases (each column $i, i \in \{1, \dots, 6\}$, shows the profit of producer i for the respective influence coefficient while the other producers remain stuck to their consistent conjectures). Thus, the results from Theorems 3.6 and 3.8 can hold true for a wider set of functions than the ones satisfying the theorems' requirements.

References

1. Bulavsky, V.A.: Structure of demand and equilibrium in a model of oligopoly. Econ. Math. Methods (Ekonomika i Matematicheskie Metody) **33**, 112–134 (1997). In Russian
2. Kalashnykova, N.I., Bulavsky, V.A., Kalashnikov, V.V.: Consistent conjectures as optimal nash strategies in the upper level game. ICIC Express Lett. **6**(4), 965–970 (2012)
3. Kalashnikov, V.V., Bulavsky, V.A., Kalashnykova, N.I., López-Ramos, F.: Consistent conjectures are optimal Cournot-Nash strategies in the meta-game. Optimization **66**(12), 2007–2024 (2017)
4. Kalashnikov, V.V., Bulavsky, V.A., Kalashnykova, N.I., Castillo-Pérez, F.J.: Mixed oligopoly with consistent conjectures. Eur. J. Oper. Res. **210**(3), 729–735 (2011)

5. Liu, Y.F., Ni, Y.X., Wu, F.F., Cai, B.: Existence and uniqueness of consistent conjectural variation equilibrium in electricity markets. Int. J. Electr. Power Energy Syst. **29**(6), 455–461 (2007)
6. Isac, G., Bulavsky, V.A., Kalashnikov, V.V.: Complementarity, Equilibrium, Efficiency and Economics. Kluwer Academic Publishers, Dordrecht (2002)

Chapter 4
Bilevel Tolls Optimization Problem with Quadratic Costs

4.1 The Bilevel Tolls Optimization Problem's Formulation

As usual, we formulate the Tolls Optimization Problem (TOP) as a single-leader-multi-follower game that occurs in a multi-commodity highway network. The usual parameters for this formulation are the following:

- The multi-commodity network is defined by a set of directed arcs $A=\{1, 2, \ldots, M\}$ representing the roads, a set of nodes $N = \{1, 2, \ldots, \eta\}$ selected as the origins, transit points, and final destinations, as well as a set of commodities $K = \{1, 2, \ldots, \kappa\}$ reflecting the groups of drivers sharing the same origins and destinations. The corresponding set powers are denoted as $|A| = M$, $|N| = \eta$, and $|K| = \kappa$.
- The set of arcs A is split into a nonempty proper subset $A_1 \subset A$ of toll arcs and its complementary subset $A_2 = A \setminus A_1$ of toll-free arcs, where $|A_1| = M_1$, $|A_1| = M_2$ and $M_1 + M_2 = M$.
- The subset $i^+ \subset A$ of arcs having the node $i \in N$ as their head and the subset $i^- \subset A$ of arcs having the node $i \in N$ as their tail.
- Every arc $a \in A$ has a fixed travel delay cost c_a and a capacity upper bound q_a.
- Each toll arc $a \in A_1$ has also a maximum toll value $t_a^{\max}$ that can be charged being still attractive enough for the drivers.
- The commodity group $k \in K$ is assigned a demand value of n_k and arranges transportation from the origin node $o(k) \in N$ to the destination node $\delta(k) \in N$.
- Thus, the demand for each commodity $k \in K$ at every node $i \in N$ of the network is given by:

$$b_i^k = \begin{cases} -n_k & \text{if } i = o(k), \\ n_k & \text{if } i = \delta(k), \\ 0 & \text{otherwise.} \end{cases} \tag{4.1}$$

The decision variables for the bilevel TOP are as follows:

J. G. Flores Muñiz et al., *Public Interest and Private Enterprize: New Developments*, Lecture Notes in Networks and Systems 138,
https://doi.org/10.1007/978-3-030-58349-1_4

- The toll values t_a for each toll-arc $a \in A_1$ which has to be decided by the leader at the upper-level. These variables are stored in the vector $t = \{t_a \mid a \in A_1\}$.
- The flows x_a^k along every arc $a \in A$ which has to be decided by the commodity $k \in K$ at the lower-level. Each commodity $k \in K$ stores its variables in the vector $x^k = \{x_a^k \mid a \in A\}$ and all the variables of the lower-level are combined into the vector $x = \{x_a^k \mid a \in A,\ k \in K\}$. We also define the vectors $x^{-k} = \{x^\ell \mid \ell \in K \setminus \{k\}\}$ and $x_a^{-k} = \{x_a^\ell \mid \ell \in K \setminus \{k\}\}$ as it is usually done in Game Theory.

In [1, 2] the delay cost for a single unit to travel through arc $a \in A$ is a constant c_a, then, the travel delay cost for each commodity $k \in K$ to travel through arc a is given by the linear function $\overline{c}_a(x_a^k) = c_a x_a^k$.

However, the travel delay cost (for a single driver) being a fixed constant value, is unrealistic in the presence of congestion. In fact, the travel delay for each driver increases with the total amount of drivers sharing the same road, so the next approach would be to use a linear approximation for the travel delay depending upon the total amount of drivers traveling along the same arc.

For our new formulation of the TOP, the following parameter is introduced:

- For a single arc $a \in A$ the travel delay generated due to the traffic congestion is given by a nonnegative factor d_a.

Then, for a single unit to travel through arc a, the travel delay cost will be given by the linear function $c_a + d_a \xi$, where ξ is the total amount of drivers traveling through the same arc a. Hence, for each arc $a \in A$, the travel delay cost for each commodity $k \in K$ traveling through arc a is given by the (nonlinear) function

$$\overline{c}_a(x_a^k; x_a^{-k}) = \left(c_a + d_a \sum_{\ell \in K} x_a^\ell \right) x_a^k = c_a x_a^k + \sum_{\ell \in K \setminus \{k\}} d_a x_a^k x_a^\ell + d_a (x_a^k)^2, \quad (4.2)$$

which leads to the quadratic structure of the lower-level objective functions (instead of the linear structure as it was assumed in all previous papers mentioned in the introduction).

The travel delay costs, the tolls values, and the congestion factors are all measured in some equivalent units for the sake of consistency.

Using the notation listed above, our extended bilevel formulation for the Tolls Optimization Problem is given by the following single-leader-multi-follower game:

$$\begin{aligned} \underset{t}{\text{Maximize}} \quad & F(t) = \sum_{k \in K} \sum_{a \in A_1} t_a x_a^k, & (4.3) \\ \text{subject to} \quad & t_a \le t_a^{\max}, \ \forall a \in A_1, & (4.4) \\ & t_a \ge 0, \ \forall a \in A_1, & (4.5) \\ & x^k \in \Psi_k(t, x^{-k}), \ \forall k \in K, & (4.6) \end{aligned}$$

where

$$\Psi_k(t, x^{-k}) = \underset{x^k}{\text{Argmin}} \quad f_k(x^k) = \sum_{a\in A_1} t_a x_a^k + \sum_{a\in A} c_a x_a^k + \sum_{\ell\in K\setminus\{k\}} \sum_{a\in A} d_a x_a^k x_a^\ell + \sum_{a\in A} d_a (x_a^k)^2, \tag{4.7}$$

$$\text{subject to} \quad \sum_{a\in i^+} x_a^k - \sum_{a\in i^-} x_a^k = b_i^k, \ \forall i \in N, \tag{4.8}$$

$$x_a^k \le q_a - \sum_{\ell\in K\setminus\{k\}} x_a^\ell, \ \forall a \in A, \tag{4.9}$$

$$x_a^k \ge 0, \ \forall a \in A. \tag{4.10}$$

Now we describe in more details the bilevel program (4.3)–(4.10):

- The leader's objective function (4.3) reflects its goal of maximizing its profit which is the sum of every toll charged times the total flow of drivers in the respective arc.
- The constraints (4.4) and (4.5) bound the tolls values to be nonnegative and not greater than the maximum tolls that can be charged to the drivers.
- The constraints (4.6) represents that the follower's variable x^k must be a solution for its respective quadratic programming problem (4.7)–(4.10) for all commodities $k \in K$. This means that the lower-level variables x^k, $k \in K$, must provide a Nash equilibrium for the non-cooperative game (4.6)–(4.10).
- The objective or payoff functions (4.7) of the followers represent their desire to minimize the total travel cost given by the tolls charges, the fixed travel delays of the roads and the travel delays caused by the traffic congestion produced by the transported commodities. It is also important to notice that in the objective function (4.7) of commodity $k \in K$, while the terms corresponding to $(x_a^k)^2$ are quadratic, the terms corresponding to $x_a^k x_a^\ell$ are in fact *linear* since the values x_a^ℓ are fixed when $\ell \neq k$.
- The constraints (4.8) are the flow conservation constraints for each commodity through the network.
- The constraints (4.9) and (4.10) are the nonnegativity and capacity constraints for each of the commodities flows.

Finally, in order to exclude trivial solutions, we make the following assumptions as in [1]:

1. There is no profitable vector that induces a negative cost cycle in the network. This condition is satisfied if all the travel costs and congestion factors are nonnegative.
2. For every commodity, there is always at least one path composed solely of toll-free arcs connecting its origin and destination nodes.

4.2 Linear-Quadratic Bilevel Program Reformulation

The bilevel formulation of the Tolls Optimization Problem described in the previous section might be inconsistent since the lower-level's Nash equilibrium required in constraints (4.6) might not be unique, thus, making the upper-level's feasible region ill-defined.

In the previous works which consider linear travel delay costs $\overline{c}_a(x_a^k) = c_a x_a^k$, this situation was handled taking advantage that the lower-level game was separable (if $d_a = 0$ for all a, then, the objective function f_k doesn't have variables from the other commodities $\ell \neq k$), which allowed them to replace the lower-level's problem with a linear programming problem where the objective function to minimize was the sum of all the followers' objective functions and the constraints were all the constraints of the followers gathered together. The latter linear programming problem always provided the Nash equilibrium for the original lower-level game since the followers' problems were separable. Moreover, if the linear lower-level still had multiple solutions, considering the optimistic approach guaranteed the leader's feasible region to be well-defined.

In our formulation (4.3)–(4.10) the followers' problems are *no longer separable*, thus, minimizing the sum of their objective functions subject to all their constraints might *not* provide the Nash equilibrium for the lower-level game (4.6)–(4.10). However, we were able to show that a solution for the lower-level problem (4.6)–(4.10) can be obtained as a solution for the following quadratic programming problem:

$$x \in \Psi(t), \tag{4.11}$$

where

$$\Psi(t) = \operatorname*{Argmin}_{x} \quad f(x) = \sum_{k\in K}\sum_{a\in A_1} t_a x_a^k + \sum_{k\in K}\sum_{a\in A} c_a x_a^k + \sum_{k\in K}\sum_{\ell\in K\setminus\{k\}}\sum_{a\in A}\frac{1}{2} d_a x_a^k x_a^\ell + \sum_{k\in K}\sum_{a\in A} d_a (x_a^k)^2, \tag{4.12}$$

$$\text{subject to} \quad \sum_{a\in i^+} x_a^k - \sum_{a\in i^-} x_a^k = b_i^k, \ \forall i \in N, \ \forall k \in K, \tag{4.13}$$

$$\sum_{k\in K} x_a^k \le q_a, \ \forall a \in A, \tag{4.14}$$

$$x_a^k \ge 0, \ \forall a \in A, \ \forall k \in K. \tag{4.15}$$

In the latter reformulation, the constraints (4.13)–(4.15) are all the constraints (4.8)–(4.10) of the followers combined together, but the objective function (4.12) is *not* the sum of all the followers' objective functions (4.7) since the coefficient corresponding to the term $x_a^k x_a^\ell$ in (4.12) is $\frac{1}{2}d_a$ while the corresponding coefficient in (4.7) is d_a whenever $k \neq \ell$.

This result is stated in the following theorem:

Theorem 4.1 *The quadratic programming problem* (4.11)–(4.15) *is convex and any of its solutions provides the Nash equilibrium for the non-cooperative game* (4.6)–(4.10).

From the proof of Theorem 4.1 we also have the following corollary:

Corollary 4.1 *If the capacity constraints* (4.9) *and* (4.14) *are removed, then, the problems* (4.6)–(4.10) *and* (4.11)–(4.15) *are equivalent.*

If all the congestion factors d_a, $a \in A$, are strictly positive, then, the quadratic programming problem (4.12)–(4.15) is strictly convex and its solution is unique; otherwise, multiple solutions might appear. If that's the case, as in the previous works, the optimistic solution is accepted in order to evaluate the leader's objective function.

Then, the bilevel single-leader-multi-follower game describing the TOP is reformulated as the following linear-quadratic bilevel programming problem:

$$\underset{t,x}{\text{Maximize}}\ F(t,x) = \sum_{k\in K}\sum_{a\in A_1} t_a x_a^k, \tag{4.16}$$

$$\text{subject to} \quad t_a \le t_a^{\max},\ \forall a \in A_1, \tag{4.17}$$

$$t_a \ge 0,\ \forall a \in A_1, \tag{4.18}$$

$$x \in \Psi(t), \tag{4.19}$$

where

$$\Psi(t) = \underset{x}{\text{Argmin}} \quad f(x) = \sum_{k\in K}\sum_{a\in A_1} t_a x_a^k + \sum_{k\in K}\sum_{a\in A} c_a x_a^k + \sum_{k\in K}\sum_{\ell\in K\setminus\{k\}}\sum_{a\in A}\frac{1}{2}d_a x_a^k x_a^\ell + \sum_{k\in K}\sum_{a\in A} d_a (x_a^k)^2, \tag{4.20}$$

$$\text{subject to} \quad \sum_{a\in i^+} x_a^k - \sum_{a\in i^-} x_a^k = b_i^k,\ \forall i \in N,\ \forall k \in K, \tag{4.21}$$

$$\sum_{k\in K} x_a^k \le q_a,\ \forall a \in A, \tag{4.22}$$

$$x_a^k \ge 0,\ \forall a \in A,\ \forall k \in K. \tag{4.23}$$

Thus, Theorem 4.2 allows us to find a solution for the bilevel single-leader-multi-follower game (4.3)–(4.10), which requires solving an equilibrium problem to compute the leader's objective function, by solving the simpler linear-quadratic bilevel programming problem (4.16)–(4.23), thus reducing the complexity to compute the leader's objective function.

4.3 The Solution Methodology

To find a solution for the Tolls Optimization Problem, we propose a heuristic algorithm processing the linear-quadratic bilevel programming problem (4.16)–(4.23). The main idea behind this heuristic is to obtain the allowable increases or decreases in the coefficients corresponding to the linear terms in the objective function such that the sets of basic variables for the solution of the original problem and the solution of the perturbed problem are the same. The procedure to compute these *allowable ranges to stay basic* (ARSB) is described in the sensitivity analysis for convex quadratic programming from [3]. An adaptation of the procedure to find the ARSB for the bilevel TOP is presented in Sect. 4.4.

Similar to the heuristic based on the allowable ranges to stay optimal (ARSO) described in [2], given an upper-level feasible solution t, we solve the lower-level quadratic problem with the aid of the Wolfe-Dual to compute the allowable ranges to stay basic $\{\Delta_a^-, \Delta_a^+\}$ for the leader's decision variables t_a, $a \in A_1$, which are the coefficients corresponding to the linear terms for the lower-level objective function. If the flow along an arc $a \in A_1$ is positive we increase the toll value t_a slightly less than the allowable increase Δ_a^+ (which may decrease the flow in the arc a but not drop to zero since the basic variables x_a^k, $k \in K$, will stay basic) in an attempt to increase the profit generated by the flow along the arc a. Otherwise, if the flow along the arc $a \in A_1$ is zero, it means that the toll assigned t_a is too high. Thus, we decrease its value by slightly more than the allowable decrease Δ_a^- so the toll can become attractive to the users again. In order to maximize the leader's objective function, we implement three different rules to decide which tolls will increase and which tolls will decrease; they are described in Sect. 4.7.

If the use of the allowable ranges to stay basic does *not* provide a better feasible solution, it could be because the current solution is a local maximum for the leader's objective function. In this case, we make use of the filled function technique first proposed in [4], then, developed in [5, 6] and widely discussed in [7, 8]. However, since the filled function method was developed for minimization, we adapted the methodology from [6] to the problem of maximization. The theory corresponding to this adaptation is presented in Sect. 4.5.

The filled function method smooths the original upper-level function (4.16) by transforming the current local maximum into a local minimum. Moreover, the modified function's structure is such that when it is maximized again, the algorithm makes a jump to a different region of the feasible set which has a good chance to lead to a better local maximum (if such exists) for the upper-level.

Once we have a new toll vector t generated by the filled function method, we proceed to maximize the leader's objective function again with the help of the allowable ranges to stay basic. When a local maximum is found again, we proceed once again with the filled function method. After several fruitless attempts in a row, say 5–10, the algorithm is stopped, and the best local maximum found is accepted as a good approximation of the global maximum solution.

4.4 The Allowable Ranges to Stay Basic Procedure

First, for the current solution t find the set $\Psi(t)$ of minimum solutions $x = \{x_a^k \mid a \in A,\ k \in K\}$ of the quadratic program (4.20)–(4.23). For each $x \in \Psi(t)$ define $x_a^0 = q_a - \sum\limits_{k \in K} x_a^k \geq 0, a \in A$.

Second, find the set $\Phi(t)$ of complementary solutions (x, y, s) for the Wolfe-Dual of the quadratic program (4.20)–(4.23), which is given by:

$$\Phi(t) = \operatorname*{Argmax}_{x,y,s} \quad \phi(x, y, s) = \sum_{k \in K} \sum_{i \in N} b_i^k y_i^k - \sum_{a \in A} q_a s_a^0$$

$$- \sum_{k \in K} \sum_{\ell \in K \setminus \{k\}} \sum_{a \in A} \frac{1}{2} d_a x_a^k x_a^\ell - \sum_{k \in K} \sum_{a \in A} d_a (x_a^k)^2, \quad (4.24)$$

$$\text{subject to} \quad \sum_{i \in a^+} y_i^k - \sum_{i \in a^-} y_i^k - s_a^0 + s_a^k - \sum_{\ell \in K \setminus \{k\}} d_a x_a^\ell$$

$$-2 d_a x_a^k = t_a + c_a, \ \forall a \in A_1, \ \forall k \in K, \quad (4.25)$$

$$\sum_{i \in a^+} y_i^k - \sum_{i \in a^-} y_i^k - s_a^0 + s_a^k - \sum_{\ell \in K \setminus \{k\}} d_a x_a^\ell$$

$$-2 d_a x_a^k = c_a, \ \forall a \in A_2, \ \forall k \in K, \quad (4.26)$$

$$x \in \Psi(t), \quad (4.27)$$

$$y_i^k \in \mathbb{R}, \ \forall i \in N, \ \forall k \in K, \quad (4.28)$$

$$s_a^k \geq 0, \ \forall a \in A, \ \forall k \in K \cup \{0\}, \quad (4.29)$$

where $y = \{y_i^k \mid i \in N,\ k \in K\}$, $s = \{s_a^k \mid a \in A,\ k \in K \cup \{0\}\}$, $a^+ = \{i \in N \mid a \in i^+\}$, $a^- = \{i \in N \mid a \in i^-\}$, $a \in A$.

Next, for the index set $\mathscr{I} = \{(a, k) \mid a \in A,\ k \in K \cup \{0\}\}$, define the partition:

$$\mathcal{B} = \{(a,k) \in \mathcal{I} \mid \sup_{x \in \Psi(t)} \{x_a^k\} > 0\}, \tag{4.30}$$

$$\mathcal{N} = \{(a,k) \in \mathcal{I} \mid \sup_{(x,y,s) \in \Phi(t)} \{s_a^k\} > 0\}, \tag{4.31}$$

$$\mathcal{T} = \mathcal{I} \setminus (\mathcal{B} \cup \mathcal{N}). \tag{4.32}$$

Then, for every $\widehat{a} \in A_1$, find the maximum $\lambda_u^{\widehat{a}}$ and minimum $\lambda_\ell^{\widehat{a}}$ of the following linear programming problem:

$$\underset{\lambda,x,y,s}{\text{Maximize/Minimize}} \quad \lambda^{\widehat{a}}(\lambda) = \lambda, \tag{4.33}$$

$$\text{subject to} \quad \sum_{a \in i^+} x_a^k - \sum_{a \in i^-} x_a^k = b_i^k, \ \forall i \in N, \ \forall k \in K, \tag{4.34}$$

$$\sum_{k \in K} x_a^k + x_a^0 = q_a, \ \forall a \in A, \tag{4.35}$$

$$\sum_{i \in \widehat{a}^+} y_i^k - \sum_{i \in \widehat{a}^-} y_i^k - s_{\widehat{a}}^0 + s_{\widehat{a}}^k - \sum_{\ell \in K \setminus \{k\}} d_{\widehat{a}} x_a^\ell$$
$$-2d_{\widehat{a}} x_a^k - \lambda = t_{\widehat{a}} + c_{\widehat{a}}, \ \forall k \in K, \tag{4.36}$$

$$\sum_{i \in a^+} y_i^k - \sum_{i \in a^-} y_i^k - s_a^0 + s_a^k - \sum_{\ell \in K \setminus \{k\}} d_a x_a^\ell$$
$$-2d_a x_a^k = t_a + c_a, \ \forall a \in A_1 \setminus \{\widehat{a}\}, \ \forall k \in K, \tag{4.37}$$

$$\sum_{i \in a^+} y_i^k - \sum_{i \in a^-} y_i^k - s_a^0 + s_a^k - \sum_{\ell \in K \setminus \{k\}} d_a x_a^\ell$$
$$-2d_a x_a^k = c_a, \ \forall a \in A_2, \ \forall k \in K, \tag{4.38}$$

$$x_a^k \geq 0, \ \forall (a,k) \in \mathcal{B}, \tag{4.39}$$

$$x_a^k = 0, \ \forall (a,k) \in \mathcal{N} \cup \mathcal{T}, \tag{4.40}$$

$$s_a^k \geq 0, \ \forall (a,k) \in \mathcal{N}, \tag{4.41}$$

$$s_a^k = 0, \ \forall (a,k) \in \mathcal{B} \cup \mathcal{T}, \tag{4.42}$$

$$-t_{\widehat{a}}^{\max} \leq \lambda \leq t_{\widehat{a}}^{\max}. \tag{4.43}$$

Finally, the allowable ranges to stay basic are given by $\Delta_a^+ = \lambda_u^a$ and $\Delta_a^- = -\lambda_\ell^a$, $a \in A_1$.

4.5 The Filled Function Adaptation

Let $u = u(t)$ be a differentiable function defined over a box $T \subset \mathbb{R}^n$ such that any local maximum point of the latter function is strictly positive.

Definition 4.1 Let $\beta > \alpha > 1$. A continuously differentiable function $Q_{t^*}(t)$ is said to be a *filled function* for the maximization problem

$$\underset{t}{\text{maximize}} \; u(t), \tag{4.44}$$

$$\text{subject to } t \in T, \tag{4.45}$$

at the point $t^* \in T$ with $u(t^*) > 0$, if:

1. t^* is a strict local minimizer of $Q_{t^*}(t)$ on T.
2. Any local maximizer $\bar{t}$ of $Q_{t^*}(t)$ on T satisfies $u(\bar{t}) > \alpha u(t^*)$ or $\bar{t}$ is a vertex of T.
3. Any local maximizer $\bar{t}$ of the optimization problem (4.44)–(4.45) with $u(\bar{t}) > \beta u(t^*)$ is a local maximizer of $Q_{t^*}(t)$ on T.
4. Any $\bar{t} \in T \setminus \{t^*\}$ with $\nabla Q_{t^*}(\bar{t}) = 0$ implies $u(\bar{t}) > \alpha u(t^*)$.

Now, to construct a filled function in the sense of Definition 4.1, define two auxiliary functions as follows. For arbitrary t and $t^* \in T$, denote $b = u(t^*) > 0$ and $v = u(t)$, define:

$$g_b(v) = \begin{cases} 1, & \text{if } v \le 0, \\ 2\dfrac{v^3}{b^3} - 3\dfrac{v^2}{b^2} + 1, & \text{if } 0 \le v \le b, \\ 0, & \text{if } v \ge b, \end{cases} \tag{4.46}$$

and

$$s_b(v) = \begin{cases} 2, & \text{if } v \le 0, \\ \dfrac{125v^3}{4b^2} - \dfrac{75v^2}{4b^2} + 2, & \text{if } 0 \le v \le \dfrac{2}{5}b, \\ 1, & \text{if } \dfrac{2}{5}b \le v \le \dfrac{4}{5}b, \\ (250 - 25b)\dfrac{v^3}{b^2} + (65b - 675)\dfrac{v^2}{b^2} & \\ \quad +(600 - 56b)\frac{v}{b} + (16b + 175), & \text{if } \dfrac{4}{5}b \le v \le b, \\ b - v, & \text{if } v \ge b. \end{cases} \tag{4.47}$$

Now, given a point $t^* \in T$ such that $u(t^*) > 0$, we define the following filled function:

$$\begin{aligned} Q_{\rho,\alpha,\beta,t^*}(t) = &- \exp(-\|t - t^*\|^2) g_{(\beta-\alpha)u(t^*)}(u(t) - \alpha u(t^*)) \\ &- \rho s_{(\beta-\alpha)u(t^*)}(u(t) - \alpha u(t^*)), \end{aligned} \tag{4.48}$$

where $\rho > 0$ is a parameter.

Based on [6] we have the following theorem:

Theorem 4.2 *Assume that the function $u(t)$ is continuously differentiable in the box $T \subset \mathbb{R}^n$ and t^* is a local maximum with $u(t^*) > 0$. Then, for any $\beta > \alpha > 1$ and $\rho > 0$, the function $Q_{\rho,\alpha,\beta,t^*}(t)$ from (4.48) is a filled function for the maximization problem (4.44)–(4.45) at the point t^*.*

Proof The proof of this theorem is similar to the one from [6]. ■

4.6 The Algorithm's Description

In this section, we describe the proposed heuristic algorithm in detail.

4.6.1 The Main Algorithm

The main procedure of the heuristic algorithm is presented below as Algorithm 1.

Algorithm 1 The Main Algorithm

Step 0: Set $m = 1$, $t_a^m = 0$, $a \in A_1$, $\rho = 1$, $\alpha = 1.5$, $\beta = 2$, $m_{FF} = 0$, $\widetilde{m} = 1$ and $FF_{max} = 5$.

Step 1: Update the current iteration m with the *ARSB version 1* algorithm.

Step 2: Set $\widehat{m} = m$ and update the current iteration m with the *ARSB version 2* algorithm. If $F(t^m, x(t^m)) > F(t^{\widehat{m}}, x(t^{\widehat{m}}))$, return to *Step 1*.

Step 3: Set $\widehat{m} = m$ and update the current iteration m with the *ARSB version 3* algorithm. If $F(t^m, x(t^m)) > F(t^{\widehat{m}}, x(t^{\widehat{m}}))$, return to *Step 1*.

Step 4: If $F(t^m, x(t^m)) > F(t^{\widetilde{m}}, x(t^{\widetilde{m}}))$, set $\alpha = 1.5$, $\beta = 2$, $m_{FF} = 0$, and $\widetilde{m} = m$. Otherwise, set $\beta = \alpha$, $\alpha = (\alpha + 1)/2$, $m_{FF} = m_{FF} + 1$, and $m = \widetilde{m}$.

Step 5: If $m_{FF} \leq FF_{max}$, update the current iteration m with the *FF method* algorithm using the current values of the parameters ρ, α, β, and return to *Step 1*.

Step 6: Return the toll vector t^m as an approximate maximum solution.

In *Step 0* of the main algorithm, we start by defining the initial solution with all tolls set to zero, the initial iteration counter $m = 1$ and the initial values for the parameters ρ, α and β required for the *FF method* algorithm. We also define another counter m_{FF} to track how many times in a row we have applied the *FF method* algorithm without improving the best solution and FF_{max} is the maximum number of fruitless attempts in a row allowed. The iteration where the best solution was found is stored in $\widetilde{m}$, so we can retrieve it when needed.

In *Step 1*, we update the current best solution t^m with the first version of the algorithms based on the allowable ranges to stay basic (which are described later).

When making *Step 2* and *Step 3*, we update the current best solution t^m with the *ARSB version 2* and *ARSB version 3* algorithms, respectively. In both cases, if a better solution is found, we try to update the solution again with the *ARSB version 1* algorithm, otherwise, we proceed to the next step. The three algorithms based on the allowable ranges to stay basic only update a previous iteration if the objective function is improved and they stop when they can't find a better solution.

In *Step 4*, we check if the current solution t^m is better than the up-to-now best solution $t^{\widetilde{m}}$. If so, the parameters α and β, and the counter m_{FF} (corresponding to the filled function algorithm used in *Step 5*) are reset to its original values and the iteration of the new best current solution is stored ($\widetilde{m} = m$). Otherwise, the parameters α and β are adjusted, the filled function counter m_{FF} increases by 1, and the iteration from the best current solution is restored ($m = \widetilde{m}$).

In *Step 5*, we update the current best solution t^m with the *FF method* algorithm and return to *Step 1* even if the new solution is worse, since the objective is to jump to another region of the feasible set (this is the reason why we need to store the iteration corresponding to the best current solution found). However, if the filled function algorithm does not find a better solution in a while (when $m_{FF} > FF_{max}$), then, the last solution found t^m (which is the best current solution found) is accepted as a good approximation of the global maximum solution (*Step 6*).

It is also important to mention that in order to compute the objective function value of a solution t^m, we first have to solve the quadratic program (4.20)–(4.23) to find the (optimistic) solution $x(t^m)$ for the lower-level and, then, evaluate $F(t^m, x(t^m))$ in the upper-level function given by (4.16).

4.6.2 The ARSB Algorithms

The three algorithms based on the allowable ranges to stay basic are described in Algorithms 7–9.

Algorithm 2 ARSB version 1

Step 0: Set $\overline{m} = m$, $\widehat{t_a} = 0$, $a \in A_1$, $G = \emptyset$ and $\varepsilon = 1 \times 10^{-3}$.

Step 1: For the current solution t^m find the allowable ranges to stay basic $\{\Delta_a^-, \Delta_a^+ \mid a \in A_1\}$. Define the set

$$A_1^+ = \left\{ a \in A_1 \;\middle|\; \sum_{k \in K} x_a^k(t^m) > 0 \right\},$$

the values $\overline{\Delta}_a^+ = \min\{t_a^m + \Delta_a^+, t_a^{\max}\} - t_a^m$, $a \in A_1^+$, and find $e \in \underset{a \in A_1^+}{\operatorname{Argmax}} \left\{ \overline{\Delta}_a^+ \cdot \sum_{k \in K} x_a^k(t^m) \right\}$.

Step 2: If $A_1^+ = \emptyset$ or $\Delta_e^+ \le \varepsilon$, go to *Step 4*. Otherwise, set

$$t_a^{m+1} = \begin{cases} \min\{t_a^m + \Delta_a^+ - \varepsilon, t_a^{\max}\} & \text{if } a = e, \\ t_a^m & \text{if } a \ne e, \end{cases} \quad \forall a \in A_1.$$

Step 3: If $F(t^{m+1}, x(t^{m+1})) > F(t^m, x(t^m))$, set $m = m + 1$ and return to *Step 1*.

Step 4: If $F(t^m, x(t^m)) > F(t^{\overline{m}}, x(t^{\overline{m}}))$, set $\overline{m} = m$ and go to *Step 6*.

Step 5: If $m = \overline{m}$, go to *Step 14*. Otherwise, set $m = \overline{m}$ and return to *Step 1*.

Step 6: Define the set

$$E_1^+ = \left\{ a \in A_1 \;\middle|\; \sum_{k \in K} x_a^k(t^m) > 0,\ t_a^m < t_a^{\max} \right\}$$

and find $e \in \operatorname{Argmax}_{a \in E_1^+} \left\{ t_a^m \cdot \sum_{k \in K} x_a^k(t^m) \right\}$.

Step 7: If $E_1^+ = \emptyset$, go to *Step 9*. Otherwise, set

$$t_a^{m+1} = \begin{cases} t_a^{\max} & \text{if } a = e, \\ t_a^m & \text{if } a \neq e, \end{cases} \quad \forall a \in A_1,$$

$\widehat{t}_e = t_e^m$, $G = G \cup \{e\}$ and $m = m + 1$.

Step 8: If $F(t^m, x(t^m)) > F(t^{\overline{m}}, x(t^{\overline{m}}))$, set $\overline{m} = m$. Go to *Step 11*.

Step 9: Set

$$t_a^{m+1} = \begin{cases} \widehat{t}_a & \text{if } a \in G, \\ t_a^m & \text{if } a \notin G, \end{cases} \quad \forall a \in A_1,$$

$m = m + 1$, $\widehat{t}_a = 0$, $a \in A_1$, and $G = \emptyset$.

Step 10: If $F(t^m, x(t^m)) > F(t^{\overline{m}}, x(t^{\overline{m}}))$, set $\overline{m} = m$. Return to *Step 1*.

Step 11: For the current solution t^m find the allowable ranges to stay basic $\{\Delta_a^-, \Delta_a^+ \mid a \in A_1\}$. Define the set

$$A_1^+ = \left\{ a \in A_1 \;\middle|\; \sum_{k \in K} x_a^k(t^m) > 0 \right\},$$

the values $\overline{\Delta}_a^+ = \min\{t_a^m + \Delta_a^+, t_a^{\max}\} - t_a^m$, $a \in A_1^+$, and find $e \in \operatorname{Argmax}_{a \in A_1^+} \left\{ \overline{\Delta}_a^+ \cdot \sum_{k \in K} x_a^k(t^m) \right\}$.

Step 12: If $A_1^+ = \emptyset$ or $\Delta_e^+ \leq \varepsilon$, return to *Step 6*. Otherwise, set

$$t_a^{m+1} = \begin{cases} \min\{t_a^m + \Delta_a^+ - \varepsilon, t_a^{\max}\} & \text{if } a = e, \\ t_a^m & \text{if } a \neq e, \end{cases} \quad \forall a \in A_1.$$

Step 13: If $F(t^{m+1}, x(t^{m+1})) > F(t^m, x(t^m))$, set $m = m + 1$ and return to *Step 8*. Otherwise, return to *Step 6*.

Step 14: Return the last iteration m and the best solution found t^m to the main algorithm.

For this first version of the ARSB algorithms, in *Step 0*, we first store the initial iteration ($\overline{m} = m$), an auxiliary vector $\widehat{t} = \{\widehat{t_a} \mid a \in A_1\}$, an auxiliary set G, and a small tolerance value ε.

From *Step 1* to *Step 5*, we find the allowable ranges to stay basic and try to improve the current solution t^m by increasing the tolls $a \in A_1$ a little less than the allowable increase $(\Delta_a^+ - \varepsilon)$ so that the flow on the arc a don't drop to zero. The tolls are increased one by one, choosing first those with the highest expected profit increase which is given by $\overline{\Delta}_a^+ \sum_{k \in K} x_a^k(t^m)$. If an improvement can't be done with this method, we go to *Step 6*. Also, we keep track of the iteration $\overline{m}$ corresponding to the best solution that has been found.

From *Step 6* to *Step 10*, we increase the tolls to its maximum values, one by one, in each of the roads that have a nonzero flow, so that the drivers search for another path that will or will not include toll arcs. If, after a single toll increases to its maximum value, the drivers' new path includes a toll arc, then, after improving the new tolls vector with the idea of *Step 1* to *Step 5*, the profit will likely be higher. This time, the tolls are increased choosing first those that have the highest profit which is given by $t_a^m \sum_{k \in K} x_a^k(t^m)$. If there is any improvement, we proceed to *Step 1* if the current solution t^m is better than the best solution found $t^{\overline{m}}$. Otherwise, we go to *Step 11*. We continue this procedure until no improvement is done. By this time, it may happen that there are no drivers in any of the toll-arcs because of the expensive costs. If the latter occurs, we reset every toll to its value before being maximized (these values are stored in the variables $\widehat{t_a}$, $a \in G$) which should generate a better solution than the previous one. Then, we return to *Step 1* to repeat the process.

Step 11 to *Step 13* are auxiliary steps that mirror the procedure of *Step 1–Step 5* but adapted to return to *Step 6*, instead of *Step 1*, when an improvement is reached.

The final *Step 14* returns the iteration m, corresponding to the best solution t^m that was found, to be used in the main algorithm.

Algorithm 3 ARSB version 2

Step 1: For the current solution t^m, define

$$A_1^+ = \left\{ a \in A_1 \;\middle|\; \sum_{k \in K} x_a^k(t^m) > 0 \right\},$$

$M_a^+ = |A_1^+|$, and enumerate its elements $A_1^+ = \{a_1, a_2, \ldots, a_{M_1^+}\}$.

Step 2: If $A_1^+ = \emptyset$, go to *Step 6*. Otherwise, set $j = 1$.

Step 3: If $j > M_a^+$, go to *Step 6*.

Step 4: Set

$$t_a^{m+1} = \begin{cases} t_a^{\max} & \text{if } a = a_j, \\ t_a^m & \text{if } a \neq a_j, \end{cases} \quad \forall a \in A_1.$$

Step 5: If $F(t^{m+1}, x(t^{m+1})) > F(t^m, x(t^m))$, set $m = m + 1$ and return to *Step 1*. Otherwise, set $j = j + 1$ and return to *Step 3*.

Step 6: Return the last iteration m and the best solution found t^m to the main algorithm.

For the second version of the ARSB algorithms, we use the same idea from *ARSB version 1* algorithm's *Step 6* to *Step 10*. We first identify the toll-arcs with positive flows and increase their tolls to the maximum one by one, but this time, only if this move provides a better solution than the previous one. After all the arcs have been tried, we return to the main algorithm with the updated solution.

Algorithm 4 ARSB version 3

Step 1: For the current solution t^m, define

$$E_1^- = \left\{ a \in A_1 \;\middle|\; \sum_{k \in K} x_a^k(t^m) = 0,\ t_a^m > 0 \right\},$$

$M_a^- = |E_1^-|$, and enumerate its elements $E_1^- = \{a_1, a_2, \ldots, a_{M_1^-}\}$.

Step 2: If $E_1^- = \emptyset$, go to *Step 6*. Otherwise, set

$$t_a^{m+1} = \begin{cases} t_a^{\max} & \text{if } a \in E_1^-, \\ t_a^m & \text{if } a \notin E_1^-, \end{cases} \quad \forall a \in A_1,$$

$m = m + 1$ and $i = 1$.

Step 3: If $i > M_a^-$, go to *Step 6*. Otherwise, for the current solution t^m find the allowable ranges to stay basic $\{\Delta_a^-, \Delta_a^+ \mid a \in A_1\}$.
Step 4: Set

$$t_a^{m+1} = \begin{cases} \max\{t_a^m - \Delta_a^- - \varepsilon, 0\} & \text{if } a = a_i, \\ t_a^m & \text{if } a \neq a_i, \end{cases} \quad \forall a \in A_1.$$

Step 5: If $F(t^{m+1}, x(t^{m+1})) > F(t^m, x(t^m))$, set $m = m + 1$. Set $i = i + 1$ and return to *Step 3*.

Step 6: Return the last iteration m and the best solution found t^m to the main algorithm.

For the third version of the ARSB algorithms, we first identify the set E_1^- of toll-arcs with zero flow, which is the consequence of the toll being too high. Then, we maximize all these tolls (which will not change the solution since the toll will be even higher) and find the allowable ranges to stay basic. Next, we select each arc of this set E_1^- one by one to decrease its value by a little more than the allowable decrease, so that the flow becomes positive. If this procedure increases the profit, we update the iteration m and the solution t^m. Otherwise, we undo this change and try with the next arc. This procedure ends when all the arcs have been tested, then, we return to the main algorithm with the updated solution.

4.6.3 The FF Method Algorithm

Finally, we present the FF method's procedures as Algorithm 10. The auxiliary functions that will be used are the ones defined in Sect. 4.5.

Algorithm 5 The FF Method

Step 0: Define the function $u(t) = F(t, x(t))$ and set $t_0 = t^m$.

Step 1: Find a local maximum t^* of the auxiliary problem:

$$\underset{t}{\text{Maximize}}\, Q_{\rho,\alpha,\beta,t_0}(t) = -\exp(-\|t - t_0\|^2) g_{(\beta-\alpha)u(t_0)}(u(t) - \alpha u(t_0))$$

$$-\rho s_{(\beta-\alpha)u(t_0)}(u(t) - \alpha u(t_0)), \tag{4.49}$$

subject to

$$t_a \leq t_a^{\max}, \ \forall a \in A_1, \tag{4.50}$$

$$t_a \geq 0, \ \forall a \in A_1. \tag{4.51}$$

Step 2: Set $t^{m+1} = t^*$ and $m = m + 1$.

Step 3: Return the iteration m and the solution found t^m to the main algorithm.

The *FF method* is very simple but it helps the main algorithm not to get stuck at a local maximum. Moreover, the auxiliary function $Q_{\rho,\alpha,\beta,t_0}(t)$ was designed in such a way that if an interior local maximum is found, then, $u(\bar{t}) > \alpha u(t_0)$, where $\alpha > 1$.

4.7 The Algorithm's Description

In this section, we describe the proposed heuristic algorithm in detail.

4.7.1 The Main Algorithm

The main procedure of the heuristic algorithm is presented below as Algorithm 1.

Algorithm 6 The Main Algorithm

Step 0: Set $m = 1$, $t_a^m = 0$, $a \in A_1$, $\rho = 1$, $\alpha = 1.5$, $\beta = 2$, $m_{FF} = 0$, $\widetilde{m} = 1$ and $FF_{max} = 5$.

Step 1: Update the current iteration m with the *ARSB version 1* algorithm.

Step 2: Set $\widehat{m} = m$ and update the current iteration m with the *ARSB version 2* algorithm. If $F(t^m, x(t^m)) > F(t^{\widehat{m}}, x(t^{\widehat{m}}))$, return to *Step 1*.

Step 3: Set $\widehat{m} = m$ and update the current iteration m with the *ARSB version 3* algorithm. If $F(t^m, x(t^m)) > F(t^{\widehat{m}}, x(t^{\widehat{m}}))$, return to *Step 1*.

Step 4: If $F(t^m, x(t^m)) > F(t^{\widetilde{m}}, x(t^{\widetilde{m}}))$, set $\alpha = 1.5$, $\beta = 2$, $m_{FF} = 0$, and $\widetilde{m} = m$. Otherwise, set $\beta = \alpha$, $\alpha = (\alpha + 1)/2$, $m_{FF} = m_{FF} + 1$, and $m = \widetilde{m}$.

Step 5: If $m_{FF} \leq FF_{max}$, update the current iteration m with the *FF method* algorithm using the current values of the parameters ρ, α, β, and return to *Step 1*.

Step 6: Return the toll vector t^m as an approximate maximum solution.

In *Step 0* of the main algorithm, we start by defining the initial solution with all tolls set to zero, the initial iteration counter $m = 1$ and the initial values for the parameters ρ, α and β required for the *FF method* algorithm. We also define another counter m_{FF} to track how many times in a row we have applied the *FF method* algorithm without improving the best solution and FF_{max} is the maximum number of fruitless attempts in a row allowed. The iteration where the best solution was found is stored in $\widetilde{m}$, so we can retrieve it when needed.

In *Step 1*, we update the current best solution t^m with the first version of the algorithms based on the allowable ranges to stay basic (which are described later).

When making *Step 2* and *Step 3*, we update the current best solution t^m with the *ARSB version 2* and *ARSB version 3* algorithms, respectively. In both cases, if a better solution is found, we try to update the solution again with the *ARSB version 1* algorithm, otherwise, we proceed to the next step. The three algorithms based on the allowable ranges to stay basic only update a previous iteration if the objective function is improved and they stop when they can't find a better solution.

In *Step 4*, we check if the current solution t^m is better than the up-to-now best solution $t^{\widetilde{m}}$. If so, the parameters α and β, and the counter m_{FF} (corresponding to the filled function algorithm used in *Step 5*) are reset to its original values and the iteration of the new best current solution is stored ($\widetilde{m} = m$). Otherwise, the parameters α and β are adjusted, the filled function counter m_{FF} increases by 1, and the iteration from the best current solution is restored ($m = \widetilde{m}$).

In *Step 5*, we update the current best solution t^m with the *FF method* algorithm and return to *Step 1* even if the new solution is worse, since the objective is to jump to another region of the feasible set (this is the reason why we need to store the iteration corresponding to the best current solution found). However, if the filled function algorithm does not find a better solution in a while (when $m_{FF} > FF_{max}$), then, the last solution found t^m (which is the best current solution found) is accepted as a good approximation of the global maximum solution (*Step 6*).

It is also important to mention that in order to compute the objective function value of a solution t^m, we first have to solve the quadratic program (4.20)–(4.23) to find the (optimistic) solution $x(t^m)$ for the lower-level and, then, evaluate $F(t^m, x(t^m))$ in the upper-level function given by (4.16).

4.7.2 The ARSB Algorithms

The three algorithms based on the allowable ranges to stay basic are described in Algorithms 7–9.

Algorithm 7 ARSB version 1

Step 0: Set $\overline{m} = m$, $\widehat{t_a} = 0$, $a \in A_1$, $G = \emptyset$ and $\varepsilon = 1 \times 10^{-3}$.

Step 1: For the current solution t^m find the allowable ranges to stay basic $\{\Delta_a^-, \Delta_a^+ \mid a \in A_1\}$. Define the set

$$A_1^+ = \left\{ a \in A_1 \;\middle|\; \sum_{k \in K} x_a^k(t^m) > 0 \right\},$$

the values $\overline{\Delta}_a^+ = \min\{t_a^m + \Delta_a^+, t_a^{\max}\} - t_a^m$, $a \in A_1^+$, and find $e \in \underset{a \in A_1^+}{\text{Argmax}} \left\{ \overline{\Delta}_a^+ \cdot \sum_{k \in K} x_a^k(t^m) \right\}$.

Step 2: If $A_1^+ = \emptyset$ or $\Delta_e^+ \leq \varepsilon$, go to *Step 4*. Otherwise, set

$$t_a^{m+1} = \begin{cases} \min\{t_a^m + \Delta_a^+ - \varepsilon, t_a^{\max}\} & \text{if } a = e, \\ t_a^m & \text{if } a \neq e, \end{cases} \quad \forall a \in A_1.$$

Step 3: If $F(t^{m+1}, x(t^{m+1})) > F(t^m, x(t^m))$, set $m = m + 1$ and return to *Step 1*.

Step 4: If $F(t^m, x(t^m)) > F(t^{\overline{m}}, x(t^{\overline{m}}))$, set $\overline{m} = m$ and go to *Step 6*.

Step 5: If $m = \overline{m}$, go to *Step 14*. Otherwise, set $m = \overline{m}$ and return to *Step 1*.

Step 6: Define the set

$$E_1^+ = \left\{ a \in A_1 \;\middle|\; \sum_{k \in K} x_a^k(t^m) > 0,\ t_a^m < t_a^{\max} \right\}$$

and find $e \in \operatorname*{Argmax}_{a \in E_1^+} \left\{ t_a^m \cdot \sum_{k \in K} x_a^k(t^m) \right\}$.

Step 7: If $E_1^+ = \emptyset$, go to *Step 9*. Otherwise, set

$$t_a^{m+1} = \begin{cases} t_a^{\max} & \text{if } a = e, \\ t_a^m & \text{if } a \neq e, \end{cases} \quad \forall a \in A_1,$$

$\widehat{t_e} = t_e^m$, $G = G \cup \{e\}$ and $m = m + 1$.

Step 8: If $F(t^m, x(t^m)) > F(t^{\overline{m}}, x(t^{\overline{m}}))$, set $\overline{m} = m$. Go to *Step 11*.

Step 9: Set

$$t_a^{m+1} = \begin{cases} \widehat{t_a} & \text{if } a \in G, \\ t_a^m & \text{if } a \notin G, \end{cases} \quad \forall a \in A_1,$$

$m = m + 1$, $\widehat{t_a} = 0$, $a \in A_1$, and $G = \emptyset$.

Step 10: If $F(t^m, x(t^m)) > F(t^{\overline{m}}, x(t^{\overline{m}}))$, set $\overline{m} = m$. Return to *Step 1*.

Step 11: For the current solution t^m find the allowable ranges to stay basic $\{\Delta_a^-, \Delta_a^+ \mid a \in A_1\}$. Define the set

$$A_1^+ = \left\{ a \in A_1 \;\middle|\; \sum_{k \in K} x_a^k(t^m) > 0 \right\},$$

the values $\overline{\Delta_a^+} = \min\{t_a^m + \Delta_a^+, t_a^{\max}\} - t_a^m$, $a \in A_1^+$, and find $e \in \operatorname*{Argmax}_{a \in A_1^+} \left\{ \overline{\Delta_a^+} \cdot \sum_{k \in K} x_a^k(t^m) \right\}$.

Step 12: If $A_1^+ = \emptyset$ or $\Delta_e^+ \leq \varepsilon$, return to *Step 6*. Otherwise, set

$$t_a^{m+1} = \begin{cases} \min\{t_a^m + \Delta_a^+ - \varepsilon, t_a^{\max}\} & \text{if } a = e, \\ t_a^m & \text{if } a \neq e, \end{cases} \quad \forall a \in A_1.$$

Step 13: If $F(t^{m+1}, x(t^{m+1})) > F(t^m, x(t^m))$, set $m = m + 1$ and return to *Step 8*. Otherwise, return to *Step 6*.

Step 14: Return the last iteration m and the best solution found t^m to the main algorithm.

For this first version of the ARSB algorithms, in *Step 0*, we first store the initial iteration ($\overline{m} = m$), an auxiliary vector $\widehat{t} = \{\widehat{t_a} \mid a \in A_1\}$, an auxiliary set G, and a small tolerance value ε.

From *Step 1* to *Step 5*, we find the allowable ranges to stay basic and try to improve the current solution t^m by increasing the tolls $a \in A_1$ a little less than the allowable increase $(\Delta_a^+ - \varepsilon)$ so that the flow on the arc a doesn't drop to zero. The tolls are increased one by one, choosing first those with the highest expected profit increase which is given by $\overline{\Delta}_a^+ \sum_{k\in K} x_a^k(t^m)$. If an improvement can't be done with this method, we go to *Step 6*. Also, we keep track of the iteration $\overline{m}$ corresponding to the best solution that has been found.

From *Step 6* to *Step 10*, we increase the tolls to its maximum values, one by one, in each of the roads that have a nonzero flow, so that the drivers search for another path that will or will not include toll arcs. If, after a single toll increases to its maximum value, the drivers' new path includes a toll arc, then, after improving the new tolls vector with the idea of *Step 1* to *Step 5*, the profit will likely be higher. This time, the tolls are increased choosing first those that have the highest profit which is given by $t_a^m \sum_{k\in K} x_a^k(t^m)$. If there is any improvement, we proceed to *Step 1* if the current solution t^m is better than the best solution found $t^{\overline{m}}$. Otherwise, we go to *Step 11*. We continue this procedure until no improvement is done. By this time, it may happen that there are no drivers in any of the toll-arcs because of the expensive costs. If the latter occurs, we reset every toll to its value before being maximized (these values are stored in the variables $\widehat{t_a}$, $a \in G$) which should generate a better solution than the previous one. Then, we return to *Step 1* to repeat the process.

Step 11 to *Step 13* are auxiliary steps that mirror the procedure of *Step 1* to *Step 5* but adapted to return to *Step 6*, instead of *Step 1*, when an improvement is reached.

The final *Step 14* returns the iteration m, corresponding to the best solution t^m that was found, to be used in the main algorithm.

Algorithm 8 ARSB version 2

Step 1: For the current solution t^m, define

$$A_1^+ = \left\{ a \in A_1 \;\middle|\; \sum_{k\in K} x_a^k(t^m) > 0 \right\},$$

$M_a^+ = |A_1^+|$, and enumerate its elements $A_1^+ = \{a_1, a_2, \ldots, a_{M_1^+}\}$.

Step 2: If $A_1^+ = \emptyset$, go to *Step 6*. Otherwise, set $j = 1$.

Step 3: If $j > M_a^+$, go to *Step 6*.

Step 4: Set

$$t_a^{m+1} = \begin{cases} t_a^{\max} & \text{if } a = a_j, \\ t_a^m & \text{if } a \neq a_j, \end{cases} \quad \forall a \in A_1.$$

Step 5: If $F(t^{m+1}, x(t^{m+1})) > F(t^m, x(t^m))$, set $m = m + 1$ and return to *Step 1*. Otherwise, set $j = j + 1$ and return to *Step 3*.

Step 6: Return the last iteration m and the best solution found t^m to the main algorithm.

For the second version of the ARSB algorithms, we use the same idea from *ARSB version 1* algorithm's *Step 6* to *Step 10*. We first identify the toll-arcs with positive flows and increase their tolls to the maximum one by one, but this time, only if this move provides a better solution than the previous one. After all the arcs have been tried, we return to the main algorithm with the updated solution.

Algorithm 9 ARSB version 3

Step 1: For the current solution t^m, define

$$E_1^- = \left\{ a \in A_1 \;\middle|\; \sum_{k \in K} x_a^k(t^m) = 0,\ t_a^m > 0 \right\},$$

$M_a^- = |E_1^-|$, and enumerate its elements $E_1^- = \{a_1, a_2, \ldots, a_{M_1^-}\}$.

Step 2: If $E_1^- = \emptyset$, go to *Step 6*. Otherwise, set

$$t_a^{m+1} = \begin{cases} t_a^{\max} & \text{if } a \in E_1^-, \\ t_a^m & \text{if } a \notin E_1^-, \end{cases} \quad \forall a \in A_1,$$

$m = m + 1$ and $i = 1$.

Step 3: If $i > M_a^-$, go to *Step 6*. Otherwise, for the current solution t^m find the allowable ranges to stay basic $\{\Delta_a^-, \Delta_a^+ \mid a \in A_1\}$.
Step 4: Set

$$t_a^{m+1} = \begin{cases} \max\{t_a^m - \Delta_a^- - \varepsilon, 0\} & \text{if } a = a_i, \\ t_a^m & \text{if } a \neq a_i, \end{cases} \quad \forall a \in A_1.$$

Step 5: If $F(t^{m+1}, x(t^{m+1})) > F(t^m, x(t^m))$, set $m = m + 1$. Set $i = i + 1$ and return to *Step 3*.

Step 6: Return the last iteration m and the best solution found t^m to the main algorithm.

For the third version of the ARSB algorithms, we first identify the set E_1^- of toll-arcs with zero flow, which is the consequence of the toll being too high. Then, we maximize all these tolls (which will not change the solution since the toll will be even higher) and find the allowable ranges to stay basic. Next, we select each arc of this set E_1^- one by one to decrease its value by a little more than the allowable decrease, so that the flow becomes positive. If this procedure increases the profit, we update the iteration m and the solution t^m. Otherwise, we undo this change and try with the next arc. This procedure ends when all the arcs have been tested, then, we return to the main algorithm with the updated solution.

4.7.3 *The FF Method Algorithm*

Finally, we present the FF method's procedures as Algorithm 10. The auxiliary functions that will be used are the ones defined in Sect. 4.5.

Algorithm 10 The FF Method

Step 0: Define the function $u(t) = F(t, x(t))$ and set $t_0 = t^m$.

Step 1: Find a local maximum t^* of the auxiliary problem:

$$\underset{t}{\text{Maximize}}\, Q_{\rho,\alpha,\beta,t_0}(t) = -\exp(-\|t - t_0\|^2) g_{(\beta-\alpha)u(t_0)}(u(t) - \alpha u(t_0))$$

$$-\rho s_{(\beta-\alpha)u(t_0)}(u(t) - \alpha u(t_0)), \tag{4.52}$$

$$\text{subject to} \quad t_a \leq t_a^{\max}, \ \forall a \in A_1, \tag{4.53}$$

$$t_a \geq 0, \ \forall a \in A_1. \tag{4.54}$$

Step 2: Set $t^{m+1} = t^*$ and $m = m + 1$.

Step 3: Return the iteration m and the solution found t^m to the main algorithm.

The *FF method* is very simple but it helps the main algorithm not to get stuck at a local maximum. Moreover, the auxiliary function $Q_{\rho,\alpha,\beta,t_0}(t)$ was designed in such a way that if an interior local maximum is found, then, $u(\bar{t}) > \alpha u(t_0)$, where $\alpha > 1$.

4.8 Numerical Results

To test the efficiency of our heuristic algorithm, its performance is compared with two algorithms from [9], and the function `fmincon` from Matlab. The algorithms from [9] are the quasi-Newton algorithm and the sharpest ascent method. Although these two algorithms were designed for the linear bilevel formulation of the Tolls Optimization Problem, they can still be used for our linear-quadratic bilevel formulation. In addition, for this comparison to be as natural as possible, half of the instances are tested with the congestion coefficients near zero to mirror the linear behavior of the lower-level from the previous formulations. The function `fmincon` from Matlab was designed to solve optimization problems with nonlinear objective functions and constraints in general, so it is worthwhile to compare its performance with other algorithms. Moreover, since the quasi-Newton, sharpest ascent and `fmincon` algorithms are designed for local optimization, the filled function method was added as an extra step to all these algorithms not to get stuck at a local maximum.

The numerical experiments were run on a computer with an Intel(R) Core(TM) i7-8750H CPU @ 2.20 GHz processor and 8.00 GB of RAM memory. The codes were

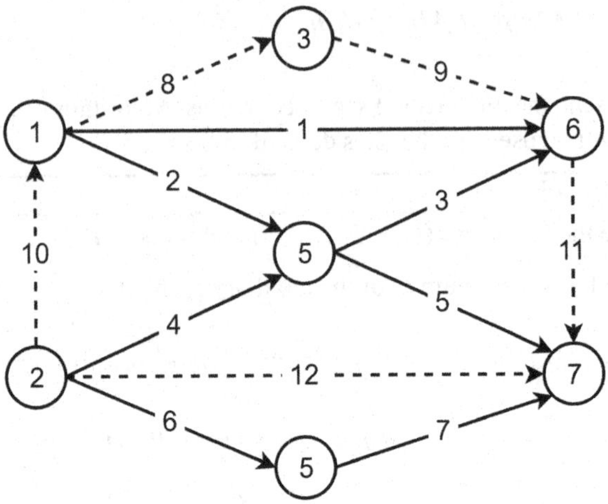

Fig. 4.1 Network 1 with 7 nodes and 12 arcs where 7 are toll arcs

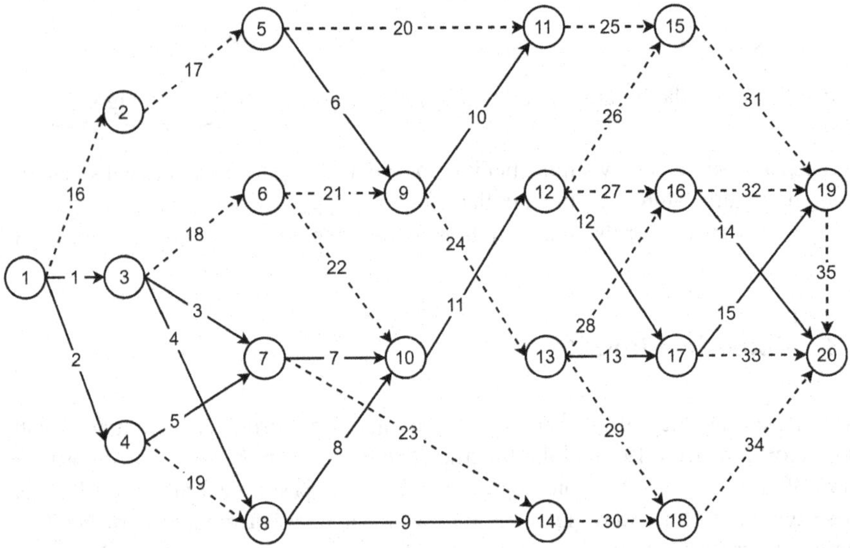

Fig. 4.2 Network 2 with 20 nodes and 35 arcs where 15 are toll arcs

compiled in Matlab R2017b making use of the linear and quadratic programming tools in the "Optimization Toolbox".

For this comparison, we used two different networks, which are shown in Figs. 4.1 and 4.2. In each of the networks, the toll arcs are represented with regular straight lines and the toll-free arcs are depicted with dotted lines.

Each of these networks was tested with 5 different instances and each of them with 2 different sets of congestion parameters; the first of them mirroring the lin-

Table 4.1 Parameters for network 1

Instance	Parameters
1	$c = (1, 2, 5, 4, 3, 3, 2, 7, 4, 3, 8, 12)$, $\Omega = \{(1, 6), (2, 7)\}$, $n = (10, 9)$.
2	$c = (3, 4, 2, 2, 3, 3, 4, 9, 9, 5, 6, 15)$, $\Omega = \{(1, 6), (2, 7)\}$, $n = (15, 5)$.
3	$c = (4, 3, 2, 1, 1, 3, 2, 5, 6, 3, 1, 5)$, $\Omega = \{(1, 6), (2, 7)\}$, $n = (5, 8)$.
4	$c = (1, 3, 1, 2, 3, 1, 1, 5, 4, 2, 4, 13)$, $\Omega = \{(1, 6), (2, 7)\}$, $n = (5, 12)$.
5	$c = (3, 4, 5, 3, 3, 6, 2, 7, 7, 8, 10, 9)$, $\Omega = \{(1, 6), (2, 7)\}$, $n = (10, 9)$.
Vector of maximum toll values that can be charged	
$t^{\max} = (20, 20, 20, 20, 20, 20, 20)$	

Table 4.2 Parameters for network 2

Instance	Parameters
1	$c = (1, 3, 4, 2, 1, 2, 2, 2, 2, 2, 4, 5, 1, 7, 9, 2, 4, 8, 7, 4, 4, 10, 12, 11, 11, 12, 9, 4, 10, 9, 13, 16, 12, 10, 13)$, $\Omega = \{(1, 15), (3, 18), (3, 20)\}$, $n = (12, 24, 30)$
2	$c = (9, 3, 7, 1, 5, 3, 4, 4, 4, 9, 1, 4, 6, 5, 6, 1, 6, 7, 7, 4, 6, 5, 2, 4, 7, 7, 8, 6, 10, 6, 5, 3, 8, 6, 11)$, $\Omega = \{(1, 15), (3, 18), (1, 20)\}$, $n = (31, 41, 120)$.
3	$c = (4, 8, 1, 7, 3, 9, 5, 5, 2, 7, 6, 6, 4, 9, 5, 5, 9, 5, 1, 4, 9, 5, 1, 4, 9, 3, 9, 1, 8, 4, 6, 3, 9, 1, 1)$, $\Omega = \{(3, 19), (3, 18), (1, 15)\}$, $n = (48, 50, 31)$.
4	$c = (1, 5, 2, 6, 3, 5, 2, 3, 7, 2, 5, 1, 6, 9, 3, 1, 3, 8, 1, 1, 10, 8, 9, 11, 6, 9, 10, 7, 7, 7, 6, 9, 10, 6, 10)$, $\Omega = \{(1, 20), (3, 18), (3, 20)\}$, $n = (84, 45, 71)$.
5	$c = (4, 3, 6, 4, 4, 3, 2, 3, 3, 2, 7, 3, 4, 5, 7, 1, 6, 4, 4, 5, 7, 3, 5, 10, 10, 9, 10, 10, 10, 7, 7, 8, 11, 10, 10)$, $\Omega = \{(1, 20), (3, 19), (3, 20)\}$, $n = (10, 6, 8)$.
Vector of maximum toll values that can be charged	
$t^{\max} = (50, 50, 50, 50, 50, 50, 50, 50, 50, 50, 50, 50, 50, 50, 50)$	

ear behavior of the Tolls Optimization Problem formulation. The sets of commodities $K = \{1, 2, \ldots, \kappa\}$ along with the origin and destination nodes $o(k)$ and $\delta(k)$ are represented together as: $\Omega = \{(o(1), \delta(1)), (o(2), \delta(2)), \ldots, (o(\kappa), \delta(\kappa))\}$. The parameters for network 1 are shown in Table 4.1 and the parameters for network 2 are shown in Table 4.2.

Networks 1 and 2 along with the instances of Table 4.1 and 4.2 were taken from [2], however, the instances of network 2 were adjusted a little to prevent trivial solutions.

The sets of congestion parameters for both networks are shown in Table 4.3.

Since the algorithms choose some values randomly, to have a better comparison, all the instances were solved 10 times by each algorithm. The initial solution was the zero vector for all the algorithms.

The results are shown in the following tables where each column is tagged according to the algorithm used. The tags are *ARSB* for the here developed algorithm, *QN* for the quasi-Newton algorithm, *SA* for the sharpest ascent algorithm, and *fmincon* for the Matlab function.

Table 4.3 Congestion parameters

Set 1 of congestion parameters for network 1
$d = (0.001, 0.001, 0.001, 0.001, 0.001, 0.001, 0.001, 0.001, 0.001, 0.001, 0.001, 0.001)$
Set 2 of congestion parameters for network 1
$d = (0.3779, 0.3815, 0.3817, 0.3799, 0.3840, 0.3772, 0.3804, 0.3796, 0.3770, 0.3804, 0.3785, 0.3771)$
Set 1 of congestion parameters for network 2
$d = (0.0001, 0.0001, 0.0001, 0.0001, 0.0001, 0.0001, 0.0001, 0.0001, 0.0001, 0.0001, 0.0001,0.0001,0.0001,0.0001,0.0001,0.0001,0.0001,0.0001,0.0001,0.0001,0.0001,0.0001, 0.0001,0.0001,0.0001,0.0001,0.0001,0.0001,0.0001,0.0001,0.0001, 0.0001,0.0001,0.0001,0.0001)$.
Set 2 of congestion parameters for network 2
$d = (0.1117, 0.1115, 0.1114, 0.1114, 0.1113, 0.1115, 0.1117, 0.1115, 0.1113, 0.1115, 0.1115, 0.1116, 0.1116, 0.1116, 0.1117, 0.1115, 0.1115, 0.1116, 0.1113, 0.1116, 0.1116, 0.1116, 0.1113, 0.1115, 0.1115, 0.1118, 0.1117, 0.1115, 0.1113, 0.1113, 0.1114, 0.1115, 0.1118, 0.1116, 0.1115)$.

Table 4.4 Best upper-level objective function found for network 1 with the congestion parameters from set 1

N1D1	ARSB	QN	SA	Fmincon	Best
1	162.74	144.74	162.74	162.74	162.74
2	274.72	264.72	264.72	274.72	274.72
3	58.85	58.85	34.97	53.82	58.85
4	171.69	171.69	131.56	95.71	171.69
5	136.74	110.25	109.80	109.90	136.74

Table 4.5 Relative error for network 1 with the congestion parameters from set 1

N1D1	ARSB (%)	QN (%)	SA (%)	Fmincon (%)
1	0.00	11.06	0.00	0.00
2	0.00	3.64	3.64	0.00
3	0.00	0.00	40.57	8.54
4	0.00	0.00	23.37	44.25
5	0.00	19.37	19.70	19.63

For the first comparison, we took the best values found by each algorithm and computed the relative error with the best value found overall. This is shown in Tables 4.4, 4.5, 4.6, 4.7, 4.8, 4.9, 4.10 and 4.11.

In Tables 4.4 and 4.11, we can see that the *best* values of the objective functions in all the instances were provided by our algorithm. In some cases, the same best solutions was also found by some of the other algorithms.

Table 4.6 Best upper-level objective function found for network 1 with the congestion parameters from set 2

N1D2	ARSB	QN	SA	Fmincon	Best
1	97.89	91.72	22.09	83.33	97.89
2	197.22	152.83	152.87	191.96	197.22
3	35.38	25.57	15.76	35.38	35.38
4	97.45	34.27	69.23	97.45	97.45
5	86.10	75.88	75.88	80.12	86.10

Table 4.7 Relative error for network 1 with the congestion parameters from set 2

N1D2	ARSB (%)	QN (%)	SA (%)	Fmincon (%)
1	0.00	6.31	77.44	14.88
2	0.00	22.51	22.49	2.67
3	0.00	27.73	55.46	0.00
4	0.00	64.83	28.96	0.00
5	0.00	11.87	11.87	6.94

Table 4.8 The best upper-level objective function value found for network 2 with the congestion parameters from set 1

N2D1	ARSB	QN	SA	Fmincon	Best
1	1760.57	1673.51	1673.28	1673.51	1760.57
2	2088.29	1532.96	1430.99	2088.29	2088.29
3	1122.81	1048.14	999.23	1122.84	1122.84
4	2166.00	2085.39	2085.26	2085.53	2166.00
5	345.91	193.96	189.60	189.97	345.91

Table 4.9 Relative errors for network 2 with the congestion parameters from set 1

N2D1	ARSB (%)	QN (%)	SA (%)	Fmincon (%)
1	0.00	4.95	4.96	4.95
2	0.00	26.59	31.48	0.00
3	0.00	6.65	11.01	0.00
4	0.00	3.72	3.73	3.72
5	0.00	43.93	45.19	45.08

Table 4.10 The best upper-level objective function value for network 2 with the congestion parameters from set 2

N2D2	ARSB	QN	SA	Fmincon	Best
1	1307.87	503.88	503.89	823.04	1307.87
2	2660.24	1730.59	1820.78	1976.87	2660.24
3	803.81	402.43	402.58	419.32	803.81
4	2183.01	274.04	455.14	1845.23	2183.01
5	262.53	110.49	110.80	225.92	262.53

Table 4.11 Relative errors for network 2 with the congestion parameters from set 2

N2D2	ARSB (%)	QN (%)	SA (%)	Fmincon (%)
1	0.00	61.47	61.47	37.07
2	0.00	34.95	31.56	25.69
3	0.00	49.93	49.92	47.83
4	0.00	87.45	79.15	15.47
5	0.00	57.91	57.80	13.95

Also, from Table 4.4, we can see that the best maximum values for the objective functions found are very similar to those found in the numerical results in [2] for the previous linear formulation of the TOP, which confirms that the linear behavior is modelled correctly, too. This also means that our quadratic formulation illustrates the continuity of the solution with respect to the parameters, i.e., when the congestion parameters tend to zero, the solution of our quadratic formulation tends to the solution of the linear setup.

In Tables 4.4 and 4.5 corresponding to the first network and the congestion coefficients near to zero, we see that in most of the instances, the other algorithms found an objective value near the best found. However, in Tables 4.6 and 4.7 corresponding to the first network and greater congestion coefficients, we see that the performance of the other algorithms is better than ours only in a few cases.

The same behavior can be seen in Tables 4.8, 4.9, 4.10 and 4.11 corresponding to network 2. When the congestion coefficients are near to zero the performance of the other algorithms is good, but when the congestion coefficients are greater, the best solution found by the other algorithms has a relative error greater than 13% compared with the best solution found by our algorithm.

However, finding better solutions is not enough for an algorithm: we also have to see if the time required to find that solution is feasible. Then, for our next comparison, we show the average solution found and the average time (in seconds) required by all the algorithms in each instance for the 10 tests. These results are shown in Tables 4.12, 4.13, 4.14, 4.15, 4.16, 4.17, 4.18 and 4.19.

In Tables 4.12, 4.13, 4.14 and 4.15 corresponding to network 1, we can see that in most of the cases the average objective function value found is very similar for

Table 4.12 The average execution time (seconds) for network 1 with the congestion parameters from set 1

N1D1	ARSB	QN	SA	Fmincon
1	19.34	4.66	6.35	10.89
2	12.06	4.68	5.48	6.22
3	12.98	7.39	4.90	8.45
4	12.82	7.14	4.19	4.69
5	14.20	4.16	4.03	5.91

Table 4.13 The average objective function value found for network 1 with the congestion parameters from set 1

N1D1	ARSB	QN	SA	Fmincon
1	160.72	126.69	138.12	162.71
2	274.72	229.06	240.07	231.60
3	51.44	48.06	34.97	53.82
4	170.96	160.48	131.43	95.71
5	130.99	109.00	109.57	109.90

Table 4.14 The average execution time (seconds) for network 1 with the congestion parameters from set 2

N1D2	ARSB	QN	SA	Fmincon
1	6.59	7.28	7.45	5.18
2	6.54	7.33	6.15	18.96
3	8.86	4.18	7.90	8.58
4	7.61	4.51	5.73	22.63
5	6.83	4.35	8.78	3.60

Table 4.15 The average objective function value found for network 1 with the congestion parameters from set 2

N1D2	ARSB	QN	SA	Fmincon
1	95.81	61.10	22.09	83.33
2	196.70	149.76	152.87	191.96
3	35.01	15.59	15.76	35.38
4	97.45	34.26	69.22	91.81
5	85.84	69.33	75.88	80.12

Table 4.16 The average execution time (seconds) for network 2 with the congestion parameters from set 1

N2D1	ARSB	QN	SA	Fmincon
1	92.20	14.07	15.59	15.92
2	84.96	18.57	8.08	33.92
3	88.16	14.32	11.68	28.27
4	110.81	10.07	7.35	19.78
5	130.88	21.92	10.54	25.55

Table 4.17 The average objective function value found for network 2 with the congestion parameters from set 1

N2D1	ARSB	QN	SA	Fmincon
1	1538.35	1279.04	1656.39	1673.23
2	2022.19	1114.04	1430.99	2004.43
3	989.02	1001.25	963.29	1122.77
4	2003.80	2081.96	2083.69	1953.31
5	288.22	138.45	158.51	189.97

Table 4.18 The average execution time (seconds) for network 2 with the congestion parameters from set 2

N2D2	ARSB	QN	SA	Fmincon
1	105.20	11.82	18.82	52.27
2	67.94	17.88	41.67	17.54
3	44.60	10.13	21.73	19.17
4	107.85	10.72	19.09	18.05
5	161.30	9.65	16.89	68.72

Table 4.19 The average objective function value found for network 2 with the congestion parameters from set 2

N2D2	ARSB	QN	SA	Fmincon
1	1270.42	503.79	503.89	823.04
2	2346.52	1229.51	1820.78	1976.87
3	739.64	401.11	402.58	419.32
4	1287.49	274.02	455.14	1845.23
5	204.53	102.38	110.79	225.92

Table 4.20 Number of final objective functions found with less than 10% relative error for network 1 with the congestion parameters in set 1

N1D1	ARSB	QN	SA	Fmincon
1	10	0	3	10
2	10	1	4	2
3	3	5	0	10
4	10	7	0	0
5	10	0	0	0

all the algorithms with set 1 of the congestion parameters, whereas, with set 2 of the congestion parameters, the average objective function value found is slightly better with our algorithm than with the other methods.

On the other hand, we also see that for set 1 of congestion parameters, the average execution time of our algorithm is almost the double of the average execution time of the other 3 algorithms. However, with set 2 of the congestion parameters, this average execution time decreases with our algorithm, while the execution time of the other 3 algorithms slightly increases.

The better performance of our algorithm with the second set of congestion parameters can be explained by the fact that it was designed specifically for the case when the lower-level is quadratic.

The same behavior can be seen in Tables 4.16, 4.17, 4.18 and 4.19 for network 2. The average objective function value is similar in most of the instances with set 1 of the congestion parameters, while with set 2 of those parameters, the average objective function value is better in almost all the cases than the same found with the other algorithms. Also, we can see that for set 1 of the congestion parameters, the average time of our algorithms is about 5–10 times higher than the time required for the other algorithms, but these times are similar for set 2 of the congestion parameters.

Thus, we can notice that the extra time taken by our algorithm results in an improvement in the objective function. However, even though our algorithm can be 10 times slower, the maximum execution time does not exceed 3 min which is still very fast.

For the last comparison, we calculate how many times each algorithm found a solution with less than 10% relative error with respect to the best solution found by all the algorithms. Next, we do the same comparison as to the average time required by each algorithm to find the solution with less than 10% relative error with respect to the best solution found by all the algorithms. These results are shown in Tables 4.20, 4.21, 4.22, 4.23, 4.24, 4.25, 4.26 and 4.27.

From Tables 4.20, 4.21, 4.22, 4.23, 4.24, 4.25, 4.26 and 4.27, we can see again that the performance of our algorithm is better for set 2 of the congestion parameters since the other algorithms most of the times cannot find a solution with less than 10% relative error. We also can see that the average time required by our algorithm is reduced drastically for almost all the instances in network 2. This means that

Table 4.21 Average time to find the first objective function with less than 10% relative error for network 1 with the congestion parameters in set 1

N1D1	ARSB	QN	SA	Fmincon
1	9.99	N/A	6.90	7.39
2	1.27	5.93	4.40	6.17
3	12.19	5.15	N/A	5.89
4	2.06	5.60	N/A	N/A
5	1.30	N/A	N/A	N/A

Table 4.22 Number of final objective functions found with less than 10% relative error for network 1 with the congestion parameters in set 2

N1D2	ARSB	QN	SA	Fmincon
1	9	1	0	0
2	10	0	0	10
3	10	0	0	10
4	10	0	0	8
5	10	0	0	10

Table 4.23 Average time to find the first objective function with less than 10% relative error for network 1 with the congestion parameters in set 2

N1D2	ARSB	QN	SA	Fmincon
1	1.03	4.64	N/A	N/A
2	2.18	N/A	N/A	2.48
3	2.40	N/A	N/A	2.56
4	1.95	N/A	N/A	9.44
5	1.14	N/A	N/A	0.42

Table 4.24 Number of final objective functions found with less than 10% relative error for network 2 with the congestion parameters in set 1

N2D1	ARSB	QN	SA	Fmincon
1	4	4	10	10
2	8	0	0	8
3	5	1	0	10
4	7	10	10	4
5	3	0	0	0

Table 4.25 Average time to find the first objective function with less than 10% relative error for network 2 with the congestion parameters in set 1

N2D1	ARSB	QN	SA	Fmincon
1	18.77	8.96	7.97	6.84
2	10.06	N/A	N/A	23.20
3	6.93	13.13	N/A	15.93
4	52.41	0.94	0.57	9.75
5	19.91	N/A	N/A	N/A

Table 4.26 Number of final objective functions found with less than 10% relative error for network 2 with the congestion parameters in set 2

N2D2	ARSB	QN	SA	Fmincon
1	10	0	0	0
2	7	0	0	0
3	8	0	0	0
4	2	0	0	0
5	1	0	0	0

Table 4.27 Average time to find the first objective function with less than 10% relative error for network 2 with the congestion parameters in set 2

N2D2	ARSB	QN	SA	Fmincon
1	28.07	N/A	N/A	N/A
2	14.34	N/A	N/A	N/A
3	4.10	N/A	N/A	N/A
4	5.30	N/A	N/A	N/A
5	129.66	N/A	N/A	N/A

our algorithm is able to find a good solution within the average time of the other algorithms, and the rest of the efforts is invested to find an even better solution.

References

1. Didi-Biha, M., Marcotte, P., Savard, G.: Path-based formulation of a bilevel toll setting problem. In: Dempe, S., Kalashnikov, V.V. (eds.) Optimization with Multi-Valued Mappings: Theory, Applications and Algorithms, pp. 29–50. Springer Science, Boston, MA (2006)
2. Kalashnikov, V.V., Herrera, R.C., Camacho, F., Kalashnykova, N.I.: A heuristic algorithm solving bilevel toll optimization problems. Int. J. Logist. Manag. **27**(1), 31–51 (2016)
3. Hadigheh, A.G., Romanko, O., Terlaky, T.: Sensitivity analysis in convex quadratic optimization: simultaneous perturbation of the objective and right-hand-side vectors. Algorithm. Oper. Res. **2**, 94–111 (2007)

4. Renpu, G.E.: A filled function method for finding a global minimizer of a function of several variables. Math. Program. **46**(1), 191–204 (1990)
5. Wan, Z., Yuan, L., Chen, J.: A filled function method for nonlinear systems of equalities and inequalities. Comput. Appl. Math. **31**(2), 391–405 (2012)
6. Wu, Z.Y., Mammadov, M., Bai, F.S., Yang, Y.J.: A filled function method for nonlinear equations. Appl. Math. Comput. **189**(2), 1196–1204 (2007)
7. Kalashnikov, V.V., Kreinovich, V., Flores-Muñiz, J.G., Kalashnykova, N.: Structure of filled functions: why gaussian and cauchy templates are most efficient. Int. J. Comb. Optim. Probl. Inform. **7**(3), 87–93 (2016)
8. Flores-Muñiz, J.G., Kalashnikov, V.V., Kreinovich, V., Kalashnykova, N.: Gaussian and cauchy functions in the filled function method - why and what next: on the example of optimizing road tolls. Acta Polytechnica Hungarica **14**(3), 237–250 (2017)
9. Kalashnikov, V.V., Camacho, F., Askin, R., Kalashnykova, N.I.: Comparison of algorithms solving a bilevel toll setting problem. Int. J. Innov. Comput. Inf. Control **6**(8), 3529–3549 (2010)

Chapter 5
Conclusions and Future Research

In Chap. 2 we presented mathematically rigorous proofs of the conjectures (cf., [1]) concerning the behavior of the semi-public company and private firm of a semi-mixed duopoly of a homogeneous good. The main difference of this work from the classical duopoly models is in the presence of one producer who maximizes not its net profit, but the convex combination of the latter with the domestic social surplus. Moreover, we not only studied the classical Cournot-Nash and perfect competition equilibriums in the model, but also the consistent conjectural variations equilibrium (CCVE) introduced and examined previously by numerous authors.

We demonstrated the existence (and in the case of the affine demand function, the uniqueness) of the CCVE and provided elements of comparative static analysis by evaluating the relationships between the equilibrium price and equilibrium production outputs of both the semi-public and private producers in the aforementioned equilibrium types.

The role of the convex combination parameter $\beta \in (0, 1]$ involved in the definition of the objective function of the semi-public (socially responsible) company was discussed and investigated. Since this parameter can be considered as reflecting the semi-public company's socialization level, we introduced a criterion to estimate its optimal value, namely, we proposed to admit the value of this parameter as desirable (optimal) if, for this parameter value, the net profits of the private firm under the CCVE and the Cournot-Nash equilibrium conditions coincide. It can be reasonable when taking into account that, with such profit equality, the socially responsible authorities need not pay any subsidies either to the private producer (to compensate its financial losses if switching from the Cournot-Nash strategy to the consistent conjectures prevailing in CCVE), or to the consumers (in the case when the equilibrium price under Cournot-Nash equilibrium turns out to be much higher than it would be in the consistent conjectural variations equilibrium, CCVE). Under the additional

J. G. Flores Muñiz et al., *Public Interest and Private Enterprize: New Developments*, Lecture Notes in Networks and Systems 138,
https://doi.org/10.1007/978-3-030-58349-1_5

assumption about the affine nature of the model's demand function, the existence of such an optimal parameter value $\overline{\beta} \in (0, 1)$ is proven.

The linearity of the demand function is a serious restriction. Hence, one of the aims for future research is to relax this condition and extend the obtained results to the semi-mixed duopoly with the demand function being not necessarily affine. The next step of our research plan is to investigate the role of the socialization level parameter in order to find its optimal value in the semi-mixed oligopoly, wherein more than one private firms compete.

Chapter 3 logically completes the previous papers [2, 3] by providing in Sect. 3.4 three results establishing (under certain rather mild conditions) the equivalence of the consistent conjectural variations equilibrium (CCVE) in an original oligopoly model to the classical Cournot-Nash equilibrium in the meta-model. The latter comprises the same agents of the original oligopoly but with their conjectures (about the possible price variations) as their strategies.

These results seem to be very interesting in two aspects. First, they could be considered as a good justification of the CVE concept as being tightly related to the classical Nash equilibrium. Second, this equivalence occurring in the oligopoly can help one develop a concept similar to the CVE but in application to other kinds of economic and financial models that lack some attributes of the oligopoly and thus do not allow one to introduce the (consistent) CVE directly. In other words, one could define the (consistent) CVE in such a model via the Nash equilibrium in the corresponding meta-model.

In our future research, we plan to implement the above-mentioned ideas, as well as extend the developed constructions to the cases of mixed oligopolies, where at least one agent endeavors to maximize not its net profit but some other function related to the social surplus. Another important extension could be related to more general economic models with not necessarily differentiable (even discontinuous) inverse demand and/or cost functions.

In Chap. 4 we propose a new formulation of the Tolls Optimization Problem with quadratic congestion terms and develop an efficient algorithm for its solution based on Sensitivity Analysis for quadratic programming. This algorithm also makes use of the filled function technique adapted for maximization in order to prevent the algorithm from getting stuck at a local maximum.

A series of numerical experiments were conducted with two different networks and two different sets of congestion parameters (one of them imitating the linear structures of previous formulations). These numerical results lead to two conclusions.

The first is that while our algorithm performs well along with the other 3 algorithms tested for small instances, our algorithm shows better results as the size of the instance increases and the congestion of the network is comparatively large (which is usually the case).

The second is that our model presents a natural extension of the linear formulation because when the congestion parameters tend to zero, the solution of our quadratic formulation tends to the solution of the linear model.

For our future research, we plan to develop efficient numerical algorithms for the cases in which the network's toll-free arcs capacity limits prevent finding the purely toll-free paths competing with the other feasible routes. These new restrictions may give birth to some theoretically novel results as well.

Finally, as an important link between this and the previous chapter, the idea inside the proof of Theorem 4.1 can be useful to reduce the many-person game from the lower level of the meta-game into a single optimization problem in future works.

References

1. Kalashnikov, V.V., Bulavsky, V.A., Kalashnykova, N.I., Watada, J., Hernández-Rodríguez, D.J.: Analysis of consistent equilibria in a mixed duopoly. J. Adv. Comput. Intell. Intell. Inform. **18**(6), 962–970 (2014)
2. Kalashnikov, V.V., Bulavsky, V.A., Kalashnykova, N.I., Castillo-Pérez, F.J.: Mixed oligopoly with consistent conjectures. Eur. J. Oper. Res. **210**(3), 729–735 (2011)
3. Kalashnikov, V.V., Bulavsky, V.A., Kalashnykova, N.I., López-Ramos, F.: Consistent conjectures are optimal Cournot-Nash strategies in the meta-game. Optimization **66**(12), 2007–2024 (2017)

Appendix A: Proofs of Results from Chapter 2

A.1 Proofs of Results from Sect. 2.2

Lemma 2.1 *Let assumptions **A1–A3** be valid. Then, for all nonnegative values of* ν_i, $i = 0, 1$, *supply values* q_i *are strictly positive (i.e.,* $q_i > 0$, $i = 0, 1$*) at any exterior equilibrium if and only if* $p > p_0$.

Proof If $p > p_0 = b_1$, then, the inequalities $p \le -\beta \nu_0 q_1 + b_0$ and $p \le b_1$, from the optimality conditions (2.11) and (2.13), respectively, never apply, which implies that no (equilibrium) value q_i, $i = 0, 1$, can vanish. Conversely, if all the equilibrium outputs are positive, i.e., $q_i > 0$, $i = 0, 1$, then, the optimality condition (2.13) directly entails $p = q_1 \nu_1 + a_1 q_1 + b_1 > b_1$. Hence, $p > p_0 = b_1$, and the proof is complete. ■

Theorem 2.1 *Under assumptions **A1–A3**, for any* $\beta \in (0, 1]$, $D \ge 0$ *and* $\nu_i \ge 0$, $i = 0, 1$, *there exists uniquely the exterior equilibrium* (p, q_0, q_1) *depending continuously upon the parameters* (D, ν_0, ν_1). *The equilibrium price p as a function of these parameters is continuously differentiable with respect to D and* ν_i, $i = 0, 1$. *Moreover* $p(D, \nu_0, \nu_1) > p_0$ *and*

$$\frac{\partial p}{\partial D} = \frac{1}{\dfrac{1}{(1-\beta)\nu_0 + a_0} + \dfrac{\nu_0 + a_0}{(1-\beta)\nu_0 + a_0}\left(\dfrac{1}{\nu_1 + a_1}\right) - G'(p)}. \tag{A.1}$$

Proof Let $\nu_0, \nu_1 \ge 0$. By using the optimality conditions (2.11) and (2.13), we can find the output volume functions $q_i = q_i(p, \nu_0, \nu_1)$, $i = 0, 1$, defined on the interval $[p_0, +\infty)$. These functions are differentiable with respect to p and ν_i, $i = 0, 1$, and they are given by:

J. G. Flores Muñiz et al., *Public Interest and Private Enterprize: New Developments*, Lecture Notes in Networks and Systems 138,
https://doi.org/10.1007/978-3-030-58349-1

$$q_0 = \frac{p - b_0}{(1-\beta)\nu_0 + a_0} + \frac{\beta\nu_0}{(1-\beta)\nu_0 + a_0}\left(\frac{p - b_1}{\nu_1 + a_1}\right), \tag{A.2}$$

$$q_1 = \frac{p - b_1}{\nu_1 + a_1}. \tag{A.3}$$

Now we introduce the following function:

$$\begin{aligned}
Q(p, \nu_0, \nu_1) =& q_0(p, \nu_0, \nu_1) + q_1(p, \nu_0, \nu_1) \\
=& \frac{p - b_0}{(1-\beta)\nu_0 + a_0} + \frac{\beta\nu_0}{(1-\beta)\nu_0 + a_0}\left(\frac{p - b_1}{\nu_1 + a_1}\right) + \frac{p - b_1}{\nu_1 + a_1} \\
=& p\left[\frac{1}{(1-\beta)\nu_0 + a_0} + \frac{\beta\nu_0}{(1-\beta)\nu_0 + a_0}\left(\frac{1}{\nu_1 + a_1}\right) + \frac{1}{\nu_1 + a_1}\right] \\
& - \left[\frac{b_0}{(1-\beta)\nu_0 + a_0} + \frac{\beta\nu_0}{(1-\beta)\nu_0 + a_0}\left(\frac{b_1}{\nu_1 + a_1}\right) + \frac{b_1}{\nu_1 + a_1}\right] \\
=& p\left[\frac{1}{(1-\beta)\nu_0 + a_0} + \left(\frac{\beta\nu_0}{(1-\beta)\nu_0 + a_0} + 1\right)\left(\frac{1}{\nu_1 + a_1}\right)\right] \\
& - \left[\frac{b_0}{(1-\beta)\nu_0 + a_0} + \left(\frac{\beta\nu_0}{(1-\beta)\nu_0 + a_0} + 1\right)\left(\frac{b_1}{\nu_1 + a_1}\right)\right] \\
=& p\left[\frac{1}{(1-\beta)\nu_0 + a_0} + \frac{\beta\nu_0 + (1-\beta)\nu_0 + a_0}{(1-\beta)\nu_0 + a_0}\left(\frac{1}{\nu_1 + a_1}\right)\right] \\
& - \left[\frac{b_0}{(1-\beta)\nu_0 + a_0} + \frac{\beta\nu_0 + (1-\beta)\nu_0 + a_0}{(1-\beta)\nu_0 + a_0}\left(\frac{b_1}{\nu_1 + a_1}\right)\right] \\
=& p\left[\frac{1}{(1-\beta)\nu_0 + a_0} + \frac{\nu_0 + a_0}{(1-\beta)\nu_0 + a_0}\left(\frac{1}{\nu_1 + a_1}\right)\right] \\
& - \left[\frac{b_0}{(1-\beta)\nu_0 + a_0} + \frac{\nu_0 + a_0}{(1-\beta)\nu_0 + a_0}\left(\frac{b_1}{\nu_1 + a_1}\right)\right].
\end{aligned} \tag{A.4}$$

As we can see from (A.4), the function Q is linear in p with positive slope. Therefore, $Q(p, \nu_0, \nu_1)$ strictly increases with respect to p, and tends to $+\infty$ when $p \to +\infty$. By assumption **A**3, one has that for all $\nu_i \geq 0$, $i = 0, 1$,

$$\begin{aligned}
Q(p_0, \nu_0, \nu_1) =& q_0(p_0, \nu_0, \nu_1) + q_1(p_0, \nu_0, \nu_1) \\
=& \frac{p_0 - b_0}{(1-\beta)\nu_0 + a_0} \leq \frac{p_0 - b_0}{a_0} < G(p_0) \leq G(p_0) + D.
\end{aligned} \tag{A.5}$$

Hence, $Q(p, \nu_0, \nu_1)$ strictly increases with respect to p, the function $G(p)$ is non-increasing by p and D is constant, so by inequality (A.5), there exists a unique value $p^* > p_0$ such that

$$Q(p^*, \nu_0, \nu_1) = G(p^*) + D. \tag{A.6}$$

For this value p^*, using (A.2) and (A.3), we compute uniquely the equilibrium output volumes $q_i^* = q_i(p^*, \nu_0, \nu_1)$, $i = 0, 1$. So we have established the existence and uniqueness of the exterior equilibrium (p^*, q_0^*, q_1^*) for any $D \geq 0$ and $\nu_i \geq 0$, $i = 0, 1$.

Now we are going to show that the equilibrium price p^* of the exterior equilibrium is differentiable with respect to the parameters (D, ν_0, ν_1). From (A.6) we get the following relationships:

$$Q(p^*, \nu_0, \nu_1) - G(p^*) - D = 0, \tag{A.7}$$

and we introduce the following function:

$$\begin{aligned} \Gamma(p^*, D, \nu_0, \nu_1) =& Q(p^*, \nu_0, \nu_1) - G(p^*) - D \\ =& p^* \left[\frac{1}{(1-\beta)\nu_0 + a_0} + \frac{\nu_0 + a_0}{(1-\beta)\nu_0 + a_0} \left(\frac{1}{\nu_1 + a_1} \right) \right] \\ & - \left[\frac{b_0}{(1-\beta)\nu_0 + a_0} + \frac{\nu_0 + a_0}{(1-\beta)\nu_0 + a_0} \left(\frac{b_1}{\nu_1 + a_1} \right) \right] \\ & - G(p^*) - D. \end{aligned} \tag{A.8}$$

Thus, we can rewrite (A.7) as a functional equation

$$\Gamma(p^*, D, \nu_0, \nu_1) = 0 \tag{A.9}$$

and compute its partial derivative with respect to p^*:

$$\begin{aligned} \frac{\partial \Gamma}{\partial p^*} =& \frac{1}{(1-\beta)\nu_0 + a_0} + \frac{\nu_0 + a_0}{(1-\beta)\nu_0 + a_0} \left(\frac{1}{\nu_1 + a_1} \right) - G'(p^*) \\ \geq & \frac{1}{(1-\beta)\nu_0 + a_0} > 0. \end{aligned} \tag{A.10}$$

From (A.10) we can see that the partial derivative of Γ with respect to p^* is positive. Because of that, Implicit Function Theorem implies that the equilibrium price p^* can be considered as a function $p^* = p^*(D, \nu_0, \nu_1)$, which is differentiable with respect to D and ν_i, $i = 0, 1$. Moreover, the partial derivative of the price p^* with respect to D can be found from the equation

$$\frac{\partial \Gamma}{\partial p^*} \frac{\partial p^*}{\partial D} + \frac{\partial \Gamma}{\partial D} = 0. \tag{A.11}$$

The latter leads to

$$\frac{\partial p^*}{\partial D} = -\frac{\dfrac{\partial \Gamma}{\partial D}}{\dfrac{\partial \Gamma}{\partial p^*}} = \frac{1}{\dfrac{1}{(1-\beta)\nu_0 + a_0} + \dfrac{\nu_0 + a_0}{(1-\beta)\nu_0 + a_0} \left(\dfrac{1}{\nu_1 + a_1} \right) - G'(p^*)}. \tag{A.12}$$

Finally, since the function p^* depends upon (D, ν_0, ν_1) and is differentiable with respect to D and ν_i, $i = 0, 1$, the functions q_i^*, $i = 0, 1$, also depend on (D, ν_0, ν_1) and are differentiable with respect to D and ν_i, $i = 0, 1$. Therefore, the equilibrium (p^*, q_0^*, q_1^*) continuously depends on the parameters (D, ν_0, ν_1). The proof of the theorem is complete ■

A.2 Proofs of Results from Sect. 2.3

Proposition 2.1 *For all $\tau \le 0$, there exists a unique solution $\nu_i = \nu_i(\tau)$, $i = 0, 1$, of system (2.20) and (2.21), which continuously depends upon τ. In addition, $\nu_i(\tau) \to 0$ whenever $\tau \to -\infty$, and $\nu_i(\tau)$ strictly grows and tends to $\nu_i(0)$ as $\tau \to 0$, $i = 0, 1$.*

Proof The variables ν_i, $i = 0, 1$, given by (2.20) and (2.21) are considered on their domains: $\nu_i \ge 0$, $a_i > 0$, $i = 0, 1$, $\beta \in (0, 1]$, and $\tau \in (-\infty, 0]$.

Substituting (2.21) in (2.20) we get the following equation:

$$
\begin{aligned}
\nu_0 &= \frac{1}{\dfrac{1}{\left(\dfrac{1}{\dfrac{1}{(1-\beta)\nu_0 + a_0} - \tau}\right) + a_1} - \tau} \\
&= \frac{1}{\dfrac{1}{\left(\dfrac{(1-\beta)\nu_0 + a_0}{1 - [(1-\beta)\nu_0 + a_0]\,\tau}\right) + a_1} - \tau} \\
&= \frac{1}{\dfrac{1}{\left(\dfrac{(1-\beta)\nu_0 + a_0}{-(1-\beta)\tau\nu_0 + (1 - a_0\tau)}\right) + a_1} - \tau} \\
&= \frac{1}{\dfrac{-(1-\beta)\tau\nu_0 + (1 - a_0\tau)}{(1-\beta)\nu_0 + a_0 + a_1\left[-(1-\beta)\tau\nu_0 + (1 - a_0\tau)\right]} - \tau} \\
&= \frac{1}{\dfrac{-(1-\beta)\tau\nu_0 + (1 - a_0\tau)}{(1-\beta)(1 - a_1\tau)\,\nu_0 + (a_0 + a_1 - a_0 a_1\tau)} - \tau} \\
&= \frac{(1-\beta)(1 - a_1\tau)\,\nu_0 + (a_0 + a_1 - a_0 a_1\tau)}{-(1-\beta)\tau\nu_0 + (1 - a_0\tau) - \left[(1-\beta)(1 - a_1\tau)\,\nu_0 + (a_0 + a_1 - a_0 a_1\tau)\right]\tau} \\
&= \frac{(1-\beta)(1 - a_1\tau)\,\nu_0 + (a_0 + a_1 - a_0 a_1\tau)}{(1-\beta)\left(-2\tau + a_1\tau^2\right)\nu_0 + \left(1 - 2a_0\tau - a_1\tau + a_0 a_1\tau^2\right)}.
\end{aligned}
\tag{A.13}
$$

Then, we can multiply (A.13) by $\left[(1-\beta)\left(-2\tau+a_1\tau^2\right)\nu_0+(1-2a_0\tau-a_1\tau+a_0a_1\tau^2)\right]$ to obtain

$$\begin{aligned}\left[(1-\beta)\left(-2\tau+a_1\tau^2\right)\nu_0+\left(1-2a_0\tau-a_1\tau+a_0a_1\tau^2\right)\right]\nu_0\\ =(1-\beta)\left(1-a_1\tau\right)\nu_0+\left(a_0+a_1-a_0a_1\tau\right).\end{aligned}\tag{A.14}$$

Move all the terms of (A.14) to the left-hand side and get

$$\begin{aligned}\left[(1-\beta)\left(-2\tau+a_1\tau^2\right)\nu_0+\left(1-2a_0\tau-a_1\tau+a_0a_1\tau^2\right)\right]\nu_0\\ -(1-\beta)\left(1-a_1\tau\right)\nu_0-\left(a_0+a_1-a_0a_1\tau\right)=0.\end{aligned}\tag{A.15}$$

By extracting a common factor from (A.15) we obtain the following quadratic equation for ν_0:

$$(1-\beta)\left(-2\tau+a_1\tau^2\right)\nu_0^2+\left(\beta-2a_0\tau-\beta a_1\tau+a_0a_1\tau^2\right)\nu_0-\left(a_0+a_1-a_0a_1\tau\right)=0.\tag{A.16}$$

Now, in order to simplify the notation, we rewrite (A.16) as follows:

$$A\nu_0^2+B\nu_0-C=0,\tag{A.17}$$

where

$$A=A(\tau)=(1-\beta)\left(-2\tau+a_1\tau^2\right)\geq 0,\tag{A.18}$$

$$B=B(\tau)=\beta-2a_0\tau-\beta a_1\tau+a_0a_1\tau^2>0,\tag{A.19}$$

$$C=C(\tau)=a_0+a_1-a_0a_1\tau>0.\tag{A.20}$$

If $\tau=0$ or $\beta=1$, then, $A=0$ and (A.17) is linear, so we can find the unique solution for ν_0 given by:

$$\nu_0(\tau)=\frac{C}{B}=\begin{cases}\dfrac{a_0+a_1}{\beta} & \text{if } \tau=0,\\[2ex] \dfrac{a_0+a_1-a_0a_1\tau}{1-2a_0\tau-a_1\tau+a_0a_1\tau^2} & \text{if } \beta=1.\end{cases}\tag{A.21}$$

If $\beta\in(0,1)$ and $\tau<0$, then, $A\neq 0$ and we can find both roots of (A.17), which are:

$$\nu_0(\tau)=\frac{-B+\sqrt{B^2+4AC}}{2A},\tag{A.22}$$

$$\nu_0(\tau)=\frac{-B-\sqrt{B^2+4AC}}{2A}.\tag{A.23}$$

However, since $\nu_0 \geq 0$, the root (A.23) is impossible; that is, (A.22) is the unique solution of (A.17).

Moreover, (A.21) and (A.22) can be combined in a single equation for all $\beta \in (0, 1]$ and $\tau \in (-\infty, 0]$ as follows:

$$\begin{aligned}\nu_0(\tau) =\nu_0 &= \frac{2C}{B + \sqrt{B^2 + 4AC}} \\ &= \frac{2(a_0+a_1-a_0a_1\tau)}{(\beta-2a_0\tau-\beta a_1\tau+a_0a_1\tau^2)+\sqrt{(\beta-2a_0\tau-\beta a_1\tau+a_0a_1\tau^2)^2+4(1-\beta)(-2\tau+a_1\tau^2)(a_0+a_1-a_0a_1\tau)}},\end{aligned} \tag{A.24}$$

where

$$B + \sqrt{B^2 + 4AC} > 0, \tag{A.25}$$

and so (A.24) is the unique solution for ν_0.

We can see that the solution (A.24) for any parameter $\beta \in (0, 1]$ satisfies the condition $\nu_0 \to 0$ as $\tau \to -\infty$. Hence, there exits a positive value $\overline{\nu_0}(\beta)$ such that $\nu_0(\tau) \leq \overline{\nu_0}(\beta)$ for all $\tau \leq 0$.

From (2.21) and (A.24), we can see that ν_1 also has a unique solution, which is given by

$$\nu_1(\tau) = \nu_1 = \frac{1}{\frac{1}{(1-\beta)\nu_0(\tau) + a_0} - \tau}. \tag{A.26}$$

For any parameter $\beta \in (0, 1]$, the conditions $\nu_1 \to 0$ as $\tau \to -\infty$ and $\nu_1(\tau) \leq a_0 + (1-\beta)\overline{\nu_0}(\beta)$ for all $\tau \leq 0$, are satisfied.

Now, it is apparent that the functions (A.18)–(A.20) are continuously differentiable with respect to τ, $\tau \in (-\infty, 0]$, and

$$A' = (1-\beta)\,(-2 + 2a_1\tau) \leq 0, \tag{A.27}$$

$$B' = -2a_0 - \beta a_1 + 2a_0a_1\tau < 0, \tag{A.28}$$

$$C' = -a_0a_1 < 0. \tag{A.29}$$

Thus, from (A.24), we have that $\nu_0(\tau)$ is continuously differentiable and

$$\begin{aligned}\nu_0' &= \frac{2C'\left(B+\sqrt{B^2+4AC}\right) - 2C\left(B' + \frac{2BB'+4A'C+4AC'}{2\sqrt{B^2+4AC}}\right)}{\left(B+\sqrt{B^2+4AC}\right)^2} \\ &= \frac{2C'\left(B+\sqrt{B^2+4AC}\right)\sqrt{B^2+4AC} - 2C\left(B'\sqrt{B^2+4AC} + \frac{2BB'+4A'C+4AC'}{2}\right)}{\left(B+\sqrt{B^2+4AC}\right)^2\sqrt{B^2+4AC}} \\ &= \frac{2C'\left(B\sqrt{B^2+4AC}+B^2+4AC\right) - 2C\left(B'\sqrt{B^2+4AC} + BB' + 2A'C + 2AC'\right)}{\left(B+\sqrt{B^2+4AC}\right)^2\sqrt{B^2+4AC}}\end{aligned}$$

$$
\begin{aligned}
&= \frac{2C'B\sqrt{B^2+4AC}+2C'B^2+4ACC'-2CB'\sqrt{B^2+4AC}-2CBB'-4A'C^2}{\left(B+\sqrt{B^2+4AC}\right)^2\sqrt{B^2+4AC}} \\
&= \frac{2\left(C'B-CB'\right)\sqrt{B^2+4AC}+2\left(C'B-CB'\right)B+4\left(AC'-A'C\right)C}{\left(B+\sqrt{B^2+4AC}\right)^2\sqrt{B^2+4AC}} \\
&= \frac{2\left(C'B-CB'\right)\left(B+\sqrt{B^2+4AC}\right)+4\left(AC'-A'C\right)C}{\left(B+\sqrt{B^2+4AC}\right)^2\sqrt{B^2+4AC}}.
\end{aligned}
\tag{A.30}
$$

Now, we estimate the values of (A.30) in order to reveal the behavior of $\nu_0(\tau)$ as the function of τ.

From (A.18)–(A.20), it is evident that the denominator of (A.30) is positive:

$$
\left(B+\sqrt{B^2+4AC}\right)^2\sqrt{B^2+4AC}>0. \tag{A.31}
$$

Thus, plugging (A.18)–(A.20) and (A.27)–(A.29) in (A.30), we can find that

$$
\begin{aligned}
C'B-CB' =&(-a_0a_1)\left(\beta-2a_0\tau-\beta a_1\tau+a_0a_1\tau^2\right) \\
&-(a_0+a_1-a_0a_1\tau)(-2a_0-\beta a_1+2a_0a_1\tau) \\
=&(-a_0a_1)\left(\beta-a_0a_1\tau^2\right)+(-a_0a_1)\left(-2a_0\tau-\beta a_1\tau+2a_0a_1\tau^2\right) \\
&-\left[(a_0+a_1)(-2a_0-\beta a_1+2a_0a_1\tau)+(-a_0a_1\tau)(-2a_0-\beta a_1+2a_0a_1\tau)\right] \\
=&a_0a_1\left(-\beta+a_0a_1\tau^2\right)-a_0a_1\left(-2a_0\tau-\beta a_1\tau+2a_0a_1\tau^2\right) \\
&+a_0a_1\left(-2a_0\tau-\beta a_1\tau+2a_0a_1\tau^2\right)+(a_0+a_1)(2a_0+\beta a_1-2a_0a_1\tau) \\
=&a_0a_1\left(-\beta+a_0a_1\tau^2\right)+(a_0+a_1)(2a_0+\beta a_1-2a_0a_1\tau) \\
=&a_0a_1(-\beta)+a_0a_1\left(a_0a_1\tau^2\right)+(a_0+a_1)(2a_0-2a_0a_1\tau)+(a_0+a_1)(\beta a_1) \\
=&-a_0(\beta a_1)+(a_0+a_1)(\beta a_1)+a_0a_1\left(a_0a_1\tau^2\right)+(a_0+a_1)(2a_0-2a_0a_1\tau) \\
=&\beta a_1^2+a_0^2a_1^2\tau^2+(a_0+a_1)(2a_0-2a_0a_1\tau)>0,
\end{aligned}
\tag{A.32}
$$

and

$$
\begin{aligned}
AC'-A'C =&(1-\beta)\left(-2\tau+a_1\tau^2\right)(-a_0a_1) \\
&-(1-\beta)(-2+2a_1\tau)(a_0+a_1-a_0a_1\tau) \\
=&(1-\beta)\left(-2\tau+2a_1\tau^2\right)(-a_0a_1)+(1-\beta)\left(-a_1\tau^2\right)(-a_0a_1) \\
&-(1-\beta)(-2+2a_1\tau)(a_0+a_1)-(1-\beta)(-2+2a_1\tau)(-a_0a_1\tau) \\
=&(1-\beta)a_0a_1^2\tau^2-(1-\beta)\left(-2\tau+2a_1\tau^2\right)(a_0a_1) \\
&+(1-\beta)\left(-2\tau+2a_1\tau^2\right)(a_0a_1)+(1-\beta)(2-2a_1\tau)(a_0+a_1) \\
=&(1-\beta)a_0a_1^2\tau^2+(1-\beta)(2-2a_1\tau)(a_0+a_1)\geq 0.
\end{aligned}
\tag{A.33}
$$

Therefore, given the values of (A.18)–(A.20), (A.25) and (A.31)–(A.33), we can conclude that

$$\begin{aligned} \nu_0' =& \frac{2\left(C'B - CB'\right)\left(B + \sqrt{B^2 + 4AC}\right) + 4\left(AC' - A'C\right)C}{\left(B + \sqrt{B^2 + 4AC}\right)^2 \sqrt{B^2 + 4AC}} \\ \geq& \frac{2\left(C'B - CB'\right)\left(B + \sqrt{B^2 + 4AC}\right)}{\left(B + \sqrt{B^2 + 4AC}\right)^2 \sqrt{B^2 + 4AC}} > 0. \end{aligned} \tag{A.34}$$

Therefore, $\nu_0(\tau)$ is strictly increasing with respect to τ, $\tau \in (-\infty, 0]$. Since the function $\nu_0 = \nu_0(\tau)$ is continuous, it tends to $\nu_0(0)$ as τ goes to 0.

Now, from (A.26) we have

$$\nu_1 = \frac{1}{\dfrac{1}{(1-\beta)\nu_0 + a_0} - \tau}. \tag{A.35}$$

Since, $\nu_0(\tau)$ is continuously differentiable with respect to τ, the same is true for $\nu_1(\tau)$, and

$$\begin{aligned} \nu_1' =& -\frac{1}{\left(\dfrac{1}{(1-\beta)\nu_0 + a_0} - \tau\right)^2}\left(-\frac{1}{[(1-\beta)\nu_0 + a_0]^2}(1-\beta)\nu_0' - 1\right) \\ =& \nu_1^2\left(\frac{(1-\beta)\nu_0'}{[(1-\beta)\nu_0 + a_0]^2} + 1\right), \end{aligned} \tag{A.36}$$

where $\nu_0' > 0$, On account of that,

$$\nu_1' = \nu_1^2\left(\frac{(1-\beta)\nu_0'}{[(1-\beta)\nu_0 + a_0]^2} + 1\right) \geq \nu_1^2 > 0. \tag{A.37}$$

Therefore, $\nu_1(\tau)$ strictly increases with respect to τ, $\tau \in (-\infty, 0]$. Since the function $\nu_1 = \nu_1(\tau)$ is continuous, it tends to $\nu_1(0)$ as τ goes to 0. The proof of the theorem is complete. ■

Theorem 2.2 *Under assumptions **A**1–**A**3, there exists the interior equilibrium.*

Proof We are going to show that there exist $\nu_i^* \geq 0, q_i^* \geq 0$, $i = 0, 1$, and $p^* > p_0$ such that the vector (p^*, q_0^*, q_1^*) is the exterior equilibrium, and the influence coefficients (ν_0^*, ν_1^*) are consistent, i.e., Eqs. (2.18) and (2.19) hold.

As it was proved in Proposition 2.1, ν_0 and ν_1 solve uniquely Eqs. (2.20) and (2.21), and continuously depend on $\tau = G'(p)$. Moreover, $G'(p)$ continuously depends on p, hence, the functions ν_0 and ν_1 are continuous with respect to p.

Recall the function (A.4) introduced when proving Theorem 2.1:

$$
\begin{aligned}
Q(p, \nu_0(p), \nu_1(p)) &= Q(p) = q_0(p, \nu_0(p), \nu_1(p)) + q_1(p, \nu_0(p), \nu_1(p)) \\
&= \frac{p - b_0}{(1-\beta)\nu_0(p) + a_0} + \frac{\beta \nu_0(p)}{(1-\beta)\nu_0(p) + a_0}\left(\frac{p - b_1}{\nu_1(p) + a_1}\right) + \frac{p - b_1}{\nu_1(p) + a_1} \\
&= p\left[\frac{1}{(1-\beta)\nu_0(p) + a_0} + \frac{\nu_0(p) + a_0}{(1-\beta)\nu_0(p) + a_0}\left(\frac{1}{\nu_1(p) + a_1}\right)\right] \\
&\quad - \left[\frac{b_0}{(1-\beta)\nu_0(p) + a_0} + \frac{\nu_0(p) + a_0}{(1-\beta)\nu_0(p) + a_0}\left(\frac{b_1}{\nu_1(p) + a_1}\right)\right].
\end{aligned}
\tag{A.38}
$$

which continuously depends on p and tends to $+\infty$ as $p \to +\infty$ since $\nu_0(p)$ and $\nu_1(p)$ are bounded. Thus, by assumption **A**3, we have that

$$
\begin{aligned}
Q(p_0) =& q_0(p_0, \nu_0(p_0), \nu_1(p_0)) + q_1(p_0, \nu_0(p_0), \nu_1(p_0)) \\
=& \frac{p_0 - b_0}{(1-\beta)\nu_0(p_0) + a_0} \le \frac{p_0 - b_0}{a_0} < G(p_0) \le G(p_0) + D.
\end{aligned}
\tag{A.39}
$$

Therefore, there exists the value $p^* > p_0$ such that

$$
Q(p^*) = G(p^*) + D. \tag{A.40}
$$

For this value p^*, we compute the influence coefficients $\nu_i^* = \nu_i(G'(p^*))$, $i = 0, 1$, using (A.24) and (A.26), as well as the output volumes $q_i^* = q_i(p^*, \nu_0^*, \nu_1^*)$, $i = 0, 1$, given by (A.2) and (A.3). Thus, ν_0^* and ν_1^* satisfy (2.18) and (2.19), whereas the vector (p^*, q_0^*, q_1^*) is the exterior equilibrium. As a consequence, the extended vector $(p^*, q_0^*, q_1^*, \nu_0^*, \nu_1^*)$ is the interior equilibrium. The proof of the theorem is complete. ■

A.3 Proofs of Results from Sect. 2.4

Corollary 2.1 *Under assumptions **A1–A3**, for all $\beta \in (0, 1]$, the demand function of type (2.22) implies the uniqueness of the interior equilibrium.*

Proof Consider an arbitrary $\beta \in (0, 1]$. Since $G'(p) = -K$, then, by Proposition 2.1, for $\tau = -K$ there exists a unique solution (ν_0^*, ν_1^*) of Eqs. (2.20) and (2.21):

$$
\nu_0^* = \frac{2(a_0 + a_1 + a_0 a_1 K)}{(\beta + 2a_0 K + \beta a_1 K + a_0 a_1 K^2) + \sqrt{(\beta + 2a_0 K + \beta a_1 K + a_0 a_1 K^2)^2 + 4(1-\beta)(2K + a_1 K^2)(a_0 + a_1 + a_0 a_1 K)}}, \tag{A.41}
$$

and

$$
\nu_1^* = \frac{1}{\dfrac{1}{(1-\beta)\nu_0^* + a_0} + K} = \frac{(1-\beta)\nu_0^* + a_0}{1 + \left[(1-\beta)\nu_0^* + a_0\right] K}. \tag{A.42}
$$

Moreover, from (2.20), we can rewrite (A.41) as follows:

$$\nu_0^* = \frac{1}{\dfrac{1}{\nu_1^* + a_1} + K}. \tag{A.43}$$

It is not difficult to see that the influence coefficients ν_0^* and ν_1^* don't depend on p, therefore, by Theorem 2.1, there exists the unique exterior equilibrium (p^*, q_0^*, q_1^*) with the influence coefficients (ν_0^*, ν_1^*). Hence, the vector

$$(p^*, q_0^*, q_1^*, \nu_0^*, \nu_1^*) = (p^*(\beta), q_0^*(\beta), q_1^*(\beta), \nu_0^*(\beta), \nu_1^*(\beta))$$

is the unique interior equilibrium for $\beta \in (0, 1]$. The proof of the corollary is complete. ■

Theorem 2.3 *For the affine demand function $G(p)$ from (2.22), the price $p^*(\beta)$, the supply outputs $q_i^*(\beta), i = 0, 1$, and the influence coefficients $\nu_i^*(\beta), i = 0, 1$, characterizing the interior equilibrium, together with total market supply $G^*(\beta) = q_0^*(\beta) + q_1^*(\beta)$, are continuously differentiable by $\beta \in (0, 1]$. Furthermore, $q_0^*(\beta)$ and $G^*(\beta)$ strictly increase, whereas $p^*(\beta)$, $\nu_0^*(\beta)$, $\nu_1^*(\beta)$ and $q_1^*(\beta)$ strictly decrease.*

Proof First, we are going to show that the functions $\nu_i^*(\beta), i = 0, 1$, are continuously differentiable and strictly decreasing with respect to β. Let us consider the functions

$$\mathscr{A} = \mathscr{A}(\beta) = (1-\beta)\left(2K + a_1 K^2\right) \geq 0, \tag{A.44}$$
$$\mathscr{B} = \mathscr{B}(\beta) = \beta + 2a_0 K + \beta a_1 K + a_0 a_1 K^2 > 0, \tag{A.45}$$
$$\mathscr{C} = \mathscr{C}(\beta) = a_0 + a_1 + a_0 a_1 K > 0, \tag{A.46}$$

which are continuously differentiable with respect to β, with

$$\mathscr{A}' = -\left(2K + a_1 K^2\right) < 0, \tag{A.47}$$
$$\mathscr{B}' = 1 + a_1 K > 0, \tag{A.48}$$
$$\mathscr{C}' = 0. \tag{A.49}$$

Using (A.44)–(A.46) we rewrite (A.41) as follows:

$$\nu_0^*(\beta) = \nu_0^* = \frac{2\mathscr{C}}{\mathscr{B} + \sqrt{\mathscr{B}^2 + 4\mathscr{A}\mathscr{C}}}, \tag{A.50}$$

where

$$\mathscr{B} + \sqrt{\mathscr{B}^2 + 4\mathscr{A}\mathscr{C}} > 0. \tag{A.51}$$

Then, $\nu_0^*(\beta)$ is continuously differentiable with respect to β and, similarly to (A.30),

$$\nu_0^{*\prime} = \frac{2\left(\mathscr{C}'\mathscr{B} - \mathscr{C}\mathscr{B}'\right)\left(\mathscr{B} + \sqrt{\mathscr{B}^2 + 4\mathscr{A}\mathscr{C}}\right) + 4\left(\mathscr{A}\mathscr{C}' - \mathscr{A}'\mathscr{C}\right)\mathscr{C}}{\left(\mathscr{B} + \sqrt{\mathscr{B}^2 + 4\mathscr{A}\mathscr{C}}\right)^2 \sqrt{\mathscr{B}^2 + 4\mathscr{A}\mathscr{C}}}. \quad \text{(A.52)}$$

Since $\mathscr{C}' = 0$, then,

$$\begin{aligned} \nu_0^{*\prime} &= \frac{2\left(-\mathscr{C}\mathscr{B}'\right)\left(\mathscr{B} + \sqrt{\mathscr{B}^2 + 4\mathscr{A}\mathscr{C}}\right) + 4\left(-\mathscr{A}'\mathscr{C}\right)\mathscr{C}}{\left(\mathscr{B} + \sqrt{\mathscr{B}^2 + 4\mathscr{A}\mathscr{C}}\right)^2 \sqrt{\mathscr{B}^2 + 4\mathscr{A}\mathscr{C}}} \\ &= \frac{-2\mathscr{C}\left[\mathscr{B}'\left(\mathscr{B} + \sqrt{\mathscr{B}^2 + 4\mathscr{A}\mathscr{C}}\right) + 2\mathscr{A}'\mathscr{C}\right]}{\left(\mathscr{B} + \sqrt{\mathscr{B}^2 + 4\mathscr{A}\mathscr{C}}\right)^2 \sqrt{\mathscr{B}^2 + 4\mathscr{A}\mathscr{C}}}. \end{aligned} \quad \text{(A.53)}$$

Now we are going to estimate the value of (A.53) in order to describe the behavior of $\nu_0^*(\beta)$ as a function of β.

From (A.44)–(A.46), it is evident that the denominator of (A.53) is positive:

$$\left(\mathscr{B} + \sqrt{\mathscr{B}^2 + 4\mathscr{A}\mathscr{C}}\right)^2 \sqrt{\mathscr{B}^2 + 4\mathscr{A}\mathscr{C}} > 0. \quad \text{(A.54)}$$

Suppose that the numerator of (A.53) is nonnegative for some $\beta_0 \in (0, 1]$, i.e,

$$-2\mathscr{C}\left[\mathscr{B}'\left(\mathscr{B} + \sqrt{\mathscr{B}^2 + 4\mathscr{A}\mathscr{C}}\right) + 2\mathscr{A}'\mathscr{C}\right] \geq 0. \quad \text{(A.55)}$$

Since $\mathscr{C} > 0$, by (A.46), we have that

$$\mathscr{B}'\left(\mathscr{B} + \sqrt{\mathscr{B}^2 + 4\mathscr{A}\mathscr{C}}\right) + 2\mathscr{A}'\mathscr{C} \leq 0. \quad \text{(A.56)}$$

Moreover, $B' > 0$, by (A.48), therefore,

$$\sqrt{\mathscr{B}^2 + 4\mathscr{A}\mathscr{C}} \leq \frac{-2\mathscr{A}'\mathscr{C}}{\mathscr{B}'} - \mathscr{B} \quad \text{(A.57)}$$

where $\sqrt{\mathscr{B}^2 + 4\mathscr{A}\mathscr{C}} > 0$. Now squaring both sides of (A.57) we have

$$\mathscr{B}^2 + 4\mathscr{A}\mathscr{C} \leq \frac{4\mathscr{A}'^2\mathscr{C}^2}{\mathscr{B}'^2} + \frac{4\mathscr{A}'\mathscr{C}\mathscr{B}}{\mathscr{B}'} + \mathscr{B}^2. \quad \text{(A.58)}$$

Solving (A.58) for $\mathscr{A}$ we get

$$\mathscr{A} \leq \frac{\mathscr{A}'^2\mathscr{C}}{\mathscr{B}'^2} + \frac{\mathscr{A}'\mathscr{B}}{\mathscr{B}'}. \quad \text{(A.59)}$$

Multiplying both sides of (A.59) by $\mathscr{B}^2$ we deduce

$$\mathscr{A}\mathscr{B}'^2 \leq \mathscr{A}'^2\mathscr{C} + \mathscr{A}'\mathscr{B}\mathscr{B}' = \mathscr{A}'\left(\mathscr{A}'\mathscr{C} + \mathscr{B}\mathscr{B}'\right). \tag{A.60}$$

Now, we substitute the values of A and A' given by (A.44) and (A.47) in (A.60) to obtain:

$$(1-\beta)\left(2K + a_1K^2\right)\mathscr{B}'^2 \leq -\left(2K + a_1K^2\right)\left[-\left(2K + a_1K^2\right)\mathscr{C} + \mathscr{B}\mathscr{B}'\right], \tag{A.61}$$

and since $\left(2K + a_1K^2\right) > 0$, we have that

$$(1-\beta)\mathscr{B}'^2 \leq -\left[-\left(2K + a_1K^2\right)\mathscr{C} + \mathscr{B}\mathscr{B}'\right] = \left(2K + a_1K^2\right)\mathscr{C} - \mathscr{B}\mathscr{B}'. \tag{A.62}$$

The latter implies

$$(1-\beta)\mathscr{B}'^2 + \mathscr{B}\mathscr{B}' - \left(2K + a_1K^2\right)\mathscr{C} = \left[(1-\beta)\mathscr{B}' + \mathscr{B}\right]\mathscr{B}' - \left(2K + a_1K^2\right)\mathscr{C} \leq 0. \tag{A.63}$$

Plugging equations (A.45), (A.46) and (A.48) in (A.63) we yield

$$\begin{aligned}
&\left[(1-\beta)\mathscr{B}' + \mathscr{B}\right]\mathscr{B}' - \left(2K + a_1K^2\right)\mathscr{C} = \\
&= \left[(1-\beta)(1 + a_1K) + \left(\beta + 2a_0K + \beta a_1K + a_0a_1K^2\right)\right](1 + a_1K) \\
&\quad - \left(2K + a_1K^2\right)(a_0 + a_1 + a_0a_1K) \\
&= \left(1 + 2a_0K + a_1K + a_0a_1K^2\right)(1 + a_1K) \\
&\quad - (2 + a_1K)\left(a_0K + a_1K + a_0a_1K^2\right) \\
&= 1 + (a_1K) + \left(2a_0K + a_1K + a_0a_1K^2\right) + \left(2a_0K + a_1K + a_0a_1K^2\right)(a_1K) \\
&\quad - (2 + a_1K)\left(a_0K + a_1K + a_0a_1K^2\right) \\
&= 1 + 2\left(a_0K + a_1K + a_0a_1K^2\right) + \left(a_0K + a_1K + a_0a_1K^2\right)(a_1K) \\
&\quad - (2 + a_1K)\left(a_0K + a_1K + a_0a_1K^2\right) \\
&= 1 + (2 + a_1K)\left(a_0K + a_1K + a_0a_1K^2\right) \\
&\quad - (2 + a_1K)\left(a_0K + a_1K + a_0a_1K^2\right) \\
&= 1 > 0,
\end{aligned} \tag{A.64}$$

which contradicts (A.63). Hence, (A.55) cannot hold for any $\beta \in (0, 1]$, which implies

$$-2\mathscr{C}\left[\mathscr{B}'\left(\mathscr{B} + \sqrt{\mathscr{B}^2 + 4\mathscr{A}\mathscr{C}}\right) + 2\mathscr{A}'\mathscr{C}\right] < 0 \tag{A.65}$$

for all $\beta \in (0, 1]$.

Therefore, from (A.54) and (A.65), we conclude that

$$\nu_0^{*\prime} = \frac{-2\mathscr{C}\left[\mathscr{B}'\left(\mathscr{B}+\sqrt{\mathscr{B}^2+4\mathscr{A}\mathscr{C}}\right)+2\mathscr{A}'\mathscr{C}\right]}{\left(\mathscr{B}+\sqrt{\mathscr{B}^2+4\mathscr{A}\mathscr{C}}\right)^2\sqrt{\mathscr{B}^2+4\mathscr{A}\mathscr{C}}} < 0 \tag{A.66}$$

for all $\beta \in (0, 1]$. On account of the latter, $\nu_0^*(\beta)$ is continuously differentiable and strictly decreasing with respect to β, $\beta \in (0, 1]$.

From (A.42), it is clear that ν_1^* is continuously differentiable with respect to ν_0^* and, since $\nu_0^*(\beta)$, on its own, is also smooth as a function of β, then, $\nu_1^*(\beta)$, is continuously differentiable by β.

Differentiating (A.43) with respect to β we get

$$\nu_0^{*\prime} = \frac{1}{\left(\dfrac{1}{\nu_1^*+a_1}+K\right)^2}\left(\frac{1}{(\nu_1^*+a_1)^2}\nu_1^{*\prime}\right) = \nu_0^{*2}\left(\frac{\nu_1^{*\prime}}{(\nu_1^*+a_1)^2}\right) = \left(\frac{\nu_0^*}{\nu_1^*+a_1}\right)^2\nu_1^{*\prime}. \tag{A.67}$$

Since $\nu_0^{*\prime} < 0$, then, (A.67) implies that $\nu_1^{*\prime} < 0$, for all $\beta \in (0, 1]$. Thus, $\nu_1^*(\beta)$ is continuously differentiable and strictly decreasing with respect to β, $\beta \in (0, 1]$. Before continuing the proof, we are going to establish the following inequality:

$$\nu_0^* + \beta\nu_0^{*\prime} > 0. \tag{A.68}$$

Substituting (A.50) and (A.53) in (A.68), we get

$$\begin{aligned}
\nu_0^* + \beta\nu_0^{*\prime} &= \frac{2\mathscr{C}}{\mathscr{B}+\sqrt{\mathscr{B}^2+4\mathscr{A}\mathscr{C}}} + \beta\frac{-2\mathscr{C}\left[\mathscr{B}'\left(\mathscr{B}+\sqrt{\mathscr{B}^2+4\mathscr{A}\mathscr{C}}\right)+2\mathscr{A}'\mathscr{C}\right]}{\left(\mathscr{B}+\sqrt{\mathscr{B}^2+4\mathscr{A}\mathscr{C}}\right)^2\sqrt{\mathscr{B}^2+4\mathscr{A}\mathscr{C}}}\\
&= \frac{2\mathscr{C}}{\left(\mathscr{B}+\sqrt{\mathscr{B}^2+4\mathscr{A}\mathscr{C}}\right)^2\sqrt{\mathscr{B}^2+4\mathscr{A}\mathscr{C}}}\\
&\quad\cdot\left\{\left(\mathscr{B}+\sqrt{\mathscr{B}^2+4\mathscr{A}\mathscr{C}}\right)\sqrt{\mathscr{B}^2+4\mathscr{A}\mathscr{C}} - \beta\left[\mathscr{B}'\left(\mathscr{B}+\sqrt{\mathscr{B}^2+4\mathscr{A}\mathscr{C}}\right)+2\mathscr{A}'\mathscr{C}\right]\right\}\\
&= \frac{2\mathscr{C}}{\left(\mathscr{B}+\sqrt{\mathscr{B}^2+4\mathscr{A}\mathscr{C}}\right)^2\sqrt{\mathscr{B}^2+4\mathscr{A}\mathscr{C}}}\\
&\quad\cdot\left[\left(-\beta\mathscr{B}'+\sqrt{\mathscr{B}^2+4\mathscr{A}\mathscr{C}}\right)\left(\mathscr{B}+\sqrt{\mathscr{B}^2+4\mathscr{A}\mathscr{C}}\right)-2\beta\mathscr{A}'\mathscr{C}\right].
\end{aligned} \tag{A.69}$$

By (A.46), (A.47) and (A.54),

$$\frac{2\mathscr{C}}{\left(\mathscr{B}+\sqrt{\mathscr{B}^2+4\mathscr{A}\mathscr{C}}\right)^2\sqrt{\mathscr{B}^2+4\mathscr{A}\mathscr{C}}} > 0 \tag{A.70}$$

and

$$-2\beta\mathscr{A}'\mathscr{C} > 0. \tag{A.71}$$

Then, to prove inequality (A.68), it suffices to show that

$$\left(-\beta\mathscr{B}' + \sqrt{\mathscr{B}^2 + 4\mathscr{A}\mathscr{C}}\right)\left(\mathscr{B} + \sqrt{\mathscr{B}^2 + 4\mathscr{A}\mathscr{C}}\right) > 0, \tag{A.72}$$

which, by (A.51), is equivalent to show that

$$-\beta\mathscr{B}' + \sqrt{\mathscr{B}^2 + 4\mathscr{A}\mathscr{C}} > 0. \tag{A.73}$$

Suppose, on the contrary, that

$$-\beta\mathscr{B}' + \sqrt{\mathscr{B}^2 + 4\mathscr{A}\mathscr{C}} \leq 0. \tag{A.74}$$

Then,

$$\sqrt{\mathscr{B}^2 + 4\mathscr{A}\mathscr{C}} \leq \beta\mathscr{B}' \tag{A.75}$$

where $\sqrt{\mathscr{B}^2 + 4\mathscr{A}\mathscr{C}} > 0$. Hence, by squaring both sides of (A.75) we have

$$\mathscr{B}^2 + 4\mathscr{A}\mathscr{C} \leq \beta^2\mathscr{B}'^2. \tag{A.76}$$

Plugging (A.45) and (A.48) in (A.76) yields

$$\left(\beta + 2a_0K + \beta a_1K + a_0a_1K^2\right)^2 + 4\mathscr{A}\mathscr{C} \leq \beta^2\left(1 + a_1K\right)^2, \tag{A.77}$$

which implies

$$\left[(\beta + \beta a_1K) + 2a_0K + a_0a_1K^2\right]^2 + 4\mathscr{A}\mathscr{C} \leq (\beta + \beta a_1K)^2. \tag{A.78}$$

However, by (A.44) and (A.46),

$$4\mathscr{A}\mathscr{C} \geq 0, \tag{A.79}$$

that is,

$$\left[(\beta + \beta a_1K) + 2a_0K + a_0a_1K^2\right]^2 \leq (\beta + \beta a_1K)^2. \tag{A.80}$$

On the other hand,

$$2a_0K + a_0a_1K^2 > 0, \tag{A.81}$$

whence

$$(\beta + \beta a_1K) < (\beta + \beta a_1K) + 2a_0K + a_0a_1K^2 \tag{A.82}$$

where $(\beta + \beta a_1K) > 0$. Now by squaring both sides of (A.75) we have

$$(\beta + \beta a_1 K)^2 < \left[(\beta + \beta a_1 K) + 2a_0 K + a_0 a_1 K^2\right]^2. \tag{A.83}$$

Nevertheless, inequality (A.83) contradicts (A.80), which means that (A.73) must hold and thus prove (A.68).

Now, coming back to the proof of the theorem, we are going to show that the equilibrium price $p^*(\beta)$ is continuously differentiable and strictly decreasing with respect to β. Consider again the function (A.4) and by plugging it in $G(p^*) = -Kp^* + T$ get the following relationships:

$$\begin{aligned} &Q(p^*, v_0^*, v_1^*) - G(p^*) - D = \\ &= p^* \left[\frac{1}{(1-\beta)v_0^* + a_0} + \frac{v_0^* + a_0}{(1-\beta)v_0^* + a_0}\left(\frac{1}{v_1^* + a_1}\right)\right] \\ &- \left[\frac{b_0}{(1-\beta)v_0^* + a_0} + \frac{v_0^* + a_0}{(1-\beta)v_0^* + a_0}\left(\frac{b_1}{v_1^* + a_1}\right)\right] \\ &+ Kp^* - T - D = 0. \end{aligned} \tag{A.84}$$

Consider the function

$$\begin{aligned} \mathscr{F}(p^*, \beta) =& p^* \left[\frac{1}{(1-\beta)v_0^* + a_0} + \frac{v_0^* + a_0}{(1-\beta)v_0^* + a_0}\left(\frac{1}{v_1^* + a_1}\right)\right] \\ &- \left[\frac{b_0}{(1-\beta)v_0^* + a_0} + \frac{v_0^* + a_0}{(1-\beta)v_0^* + a_0}\left(\frac{b_1}{v_1^* + a_1}\right)\right] \\ &+ Kp^* - T - D, \end{aligned} \tag{A.85}$$

having in mind that v_0^* and v_1^* depend on β, but not on p^*. Now, we rewrite (A.84) using (A.85) as a functional equation:

$$\mathscr{F}(p^*, \beta) = 0. \tag{A.86}$$

Now we are in a position to estimate the value of the partial derivative of the function $\mathscr{F}(p^*, \beta)$ with respect to p^*:

$$\frac{\partial \mathscr{F}}{\partial p^*} = \frac{1}{(1-\beta)v_0^* + a_0} + \frac{v_0^* + a_0}{(1-\beta)v_0^* + a_0}\left(\frac{1}{v_1^* + a_1}\right) + K \geq K > 0. \tag{A.87}$$

We observe that the partial derivative $\mathscr{F}$ with respect to p^* is positive. Hence, by the Implicit Function Theorem, the function $p^* = p^*(\beta)$ is differentiable with respect to β, and its partial derivative with respect to β can be found from the equation

$$\frac{\partial \mathscr{F}}{\partial p^*}\frac{dp^*}{d\beta} + \frac{\partial \mathscr{F}}{\partial \beta} = 0, \tag{A.88}$$

which leads to

$$\frac{dp^*}{d\beta} = -\frac{\dfrac{\partial \mathscr{F}}{\partial \beta}}{\dfrac{\partial \mathscr{F}}{\partial p^*}}. \tag{A.89}$$

From (A.87), we have

$$\frac{\partial \mathscr{F}}{\partial p^*} > 0. \tag{A.90}$$

Therefore, to prove that p^* is strictly increasing, we have to show that

$$\frac{\partial \mathscr{F}}{\partial \beta} > 0. \tag{A.91}$$

Indeed,

$$\begin{aligned}
\frac{\partial \mathscr{F}}{\partial \beta} =& \frac{\partial}{\partial \beta}\left(Q(p^*, v_0^*, v_1^*) - G(p^*) - D\right) = \frac{\partial}{\partial \beta} Q(p^*, v_0^*, v_1^*) \\
=& \frac{\partial}{\partial \beta}\left(q_0(p^*, v_0^*, v_1^*) + q_1(p^*, v_0^*, v_1^*)\right) \\
=& \frac{\partial}{\partial \beta}\left[\frac{p^* - b_0}{(1-\beta)v_0^* + a_0} + \frac{\beta v_0^*}{(1-\beta)v_0^* + a_0}\left(\frac{p^* - b_1}{v_1^* + a_1}\right) + \frac{p^* - b_1}{v_1^* + a_1}\right] \\
=& -\frac{p^* - b_0}{\left[(1-\beta)v_0^* + a_0\right]^2}\left[-v_0^* + (1-\beta)v_0^{*\prime}\right] \\
&+ \frac{\left(v_0^* + \beta v_0^{*\prime}\right)\left[(1-\beta)v_0^* + a_0\right] - \beta v_0^*\left[-v_0^* + (1-\beta)v_0^{*\prime}\right]}{\left[(1-\beta)v_0^* + a_0\right]^2}\left(\frac{p^* - b_1}{v_1^* + a_1}\right) \\
&+ \frac{\beta v_0^*}{(1-\beta)v_0^* + a_0}\left(-\frac{p^* - b_1}{\left(v_1^* + a_1\right)^2} v_1^{*\prime}\right) + \left(-\frac{p^* - b_1}{\left(v_1^* + a_1\right)^2} v_1^{*\prime}\right) \\
=& \frac{p^* - b_0}{\left[(1-\beta)v_0^* + a_0\right]^2}\left[v_0^* + (1-\beta)\left(-v_0^{*\prime}\right)\right] \\
&+ \frac{\left(v_0^* + \beta v_0^{*\prime}\right)\left[(1-\beta)v_0^* + a_0\right] + \beta v_0^*\left[v_0^* + (1-\beta)\left(-v_0^{*\prime}\right)\right]}{\left[(1-\beta)v_0^* + a_0\right]^2}\left(\frac{p^* - b_1}{v_1^* + a_1}\right) \\
&+ \frac{\beta v_0^*}{(1-\beta)v_0^* + a_0}\left[\frac{p^* - b_1}{\left(v_1^* + a_1\right)^2}\left(-v_1^{*\prime}\right)\right] + \left[\frac{p^* - b_1}{\left(v_1^* + a_1\right)^2}\left(-v_1^{*\prime}\right)\right].
\end{aligned} \tag{A.92}$$

Given the values of a_0, a_1, b_0, b_1, β, v_0^*, v_1^*, $v_0^{*\prime}$, $v_1^{*\prime}$, p^* and Eq. (A.68), it isn't difficult to see that (A.92) is nonnegative. Moreover,

$$
\begin{aligned}
\frac{\partial \mathcal{F}}{\partial \beta} =& \frac{p^* - b_0}{\left[(1-\beta)\nu_0^* + a_0\right]^2}\left[\nu_0^* + (1-\beta)\left(-\nu_0^{*\prime}\right)\right] \\
&+ \frac{\left(\nu_0^* + \beta\nu_0^{*\prime}\right)\left[(1-\beta)\nu_0^* + a_0\right] + \beta\nu_0^*\left[\nu_0^* + (1-\beta)\left(-\nu_0^{*\prime}\right)\right]}{\left[(1-\beta)\nu_0^* + a_0\right]^2}\left(\frac{p^* - b_1}{\nu_1^* + a_1}\right) \\
&+ \frac{\beta\nu_0^*}{(1-\beta)\nu_0^* + a_0}\left[\frac{p^* - b_1}{\left(\nu_1^* + a_1\right)^2}\left(-\nu_1^{*\prime}\right)\right] + \left[\frac{p^* - b_1}{\left(\nu_1^* + a_1\right)^2}\left(-\nu_1^{*\prime}\right)\right] \\
\geq& \frac{p^* - b_1}{\left(\nu_1^* + a_1\right)^2}\left(-\nu_1^{*\prime}\right) > 0,
\end{aligned}
\tag{A.93}
$$

which proves (A.91). On account of that,

$$
\frac{dp^*}{d\beta} = -\frac{\dfrac{\partial \mathcal{F}}{\partial \beta}}{\dfrac{\partial \mathcal{F}}{\partial p^*}} < 0, \tag{A.94}
$$

where $\dfrac{\partial \mathcal{F}}{\partial \beta}$ and $\dfrac{\partial \mathcal{F}}{\partial p^*}$ are continuous with respect to β. Hence $p^*(\beta)$ is continuously differentiable and strictly decreasing with respect to β, $\beta \in (0, 1]$.

Now, since

$$
G^*(\beta) = G(p^*(\beta)) = -Kp^*(\beta) + T, \tag{A.95}
$$

and $p^*(\beta)$ is continuously differentiable and strictly decreasing with respect to β, and K and T are positive constants, then, $G^*(\beta)$ is continuously differentiable and strictly increasing with respect to β, $\beta \in (0, 1]$.

Now, we are going to show that $q_1^*(\beta)$ is continuously differentiable and strictly decreasing with respect to β. To do that, we first solve Eq. (A.84) for p^* to obtain the following equality:

$$
\begin{aligned}
p^* =& \frac{\dfrac{b_0}{(1-\beta)\nu_0^* + a_0} + \dfrac{\nu_0^* + a_0}{(1-\beta)\nu_0^* + a_0}\left(\dfrac{b_1}{\nu_1^* + a_1}\right) + T + D}{\dfrac{1}{(1-\beta)\nu_0^* + a_0} + \dfrac{\nu_0^* + a_0}{(1-\beta)\nu_0^* + a_0}\left(\dfrac{1}{\nu_1^* + a_1}\right) + K} \\
=& \frac{b_0 + \left(\nu_0^* + a_0\right)\left(\dfrac{b_1}{\nu_1^* + a_1}\right) + \left[(1-\beta)\nu_0^* + a_0\right](T + D)}{1 + \left(\nu_0^* + a_0\right)\left(\dfrac{1}{\nu_1^* + a_1}\right) + \left[(1-\beta)\nu_0^* + a_0\right]K} \\
=& \frac{(\nu_0^* + a_0)b_1 + \left(\nu_1^* + a_1\right)b_0 + \left[(1-\beta)\nu_0^* + a_0\right]\left(\nu_1^* + a_1\right)(T + D)}{\left(\nu_0^* + a_0\right) + \left(\nu_1^* + a_1\right) + \left[(1-\beta)\nu_0^* + a_0\right]\left(\nu_1^* + a_1\right)K}.
\end{aligned}
\tag{A.96}
$$

We substitute (A.96) in $q_1^* = q_1(p^*, \nu_0^*, \nu_1^*)$, to deduce

$$
\begin{aligned}
q_1^* =& \frac{p^* - b_1}{v_1^* + a_1} \\
=& \frac{\dfrac{(v_0^* + a_0)\, b_1 + (v_1^* + a_1)\, b_0 + \left[(1-\beta)v_0^* + a_0\right](v_1^* + a_1)(T + D)}{(v_0^* + a_0) + (v_1^* + a_1) + \left[(1-\beta)v_0^* + a_0\right](v_1^* + a_1) K} - b_1}{v_1^* + a_1} \\
=& \frac{(v_0^* + a_0)\, b_1 + (v_1^* + a_1)\, b_0 + \left[(1-\beta)v_0^* + a_0\right](v_1^* + a_1)(T + D)}{(v_1^* + a_1)\left\{(v_0^* + a_0) + (v_1^* + a_1) + \left[(1-\beta)v_0^* + a_0\right](v_1^* + a_1) K\right\}} \\
&- \frac{\left\{(v_0^* + a_0) + (v_1^* + a_1) + \left[(1-\beta)v_0^* + a_0\right](v_1^* + a_1) K\right\} b_1}{(v_1^* + a_1)\left\{(v_0^* + a_0) + (v_1^* + a_1) + \left[(1-\beta)v_0^* + a_0\right](v_1^* + a_1) K\right\}} \\
=& \frac{(v_1^* + a_1)\, b_0 + \left[(1-\beta)v_0^* + a_0\right](v_1^* + a_1)(T + D)}{(v_1^* + a_1)\left\{(v_0^* + a_0) + (v_1^* + a_1) + \left[(1-\beta)v_0^* + a_0\right](v_1^* + a_1) K\right\}} \\
&- \frac{(v_1^* + a_1)\, b_1 + \left[(1-\beta)v_0^* + a_0\right](v_1^* + a_1) K b_1}{(v_1^* + a_1)\left\{(v_0^* + a_0) + (v_1^* + a_1) + \left[(1-\beta)v_0^* + a_0\right](v_1^* + a_1) K\right\}} \\
=& \frac{b_0 + \left[(1-\beta)v_0^* + a_0\right](T + D)}{(v_0^* + a_0) + (v_1^* + a_1) + \left[(1-\beta)v_0^* + a_0\right](v_1^* + a_1) K} \\
&- \frac{b_1 + \left[(1-\beta)v_0^* + a_0\right] K b_1}{(v_0^* + a_0) + (v_1^* + a_1) + \left[(1-\beta)v_0^* + a_0\right](v_1^* + a_1) K} \\
=& \frac{-b_1 + b_0 + \left[(1-\beta)v_0^* + a_0\right](-K b_1 + T + D)}{(v_0^* + a_0) + (v_1^* + a_1) + \left[(1-\beta)v_0^* + a_0\right](v_1^* + a_1) K} \\
=& \frac{-(b_1 - b_0) + \left[(1-\beta)v_0^* + a_0\right](G(b_1) + D)}{(v_0^* + a_0) + (v_1^* + a_1)\left\{1 + \left[(1-\beta)v_0^* + a_0\right] K\right\}}.
\end{aligned}
\tag{A.97}
$$

By plugging (A.42) in (A.97) we have that

$$
\begin{aligned}
q_1^* =& \frac{-(b_1 - b_0) + \left[(1-\beta)v_0^* + a_0\right](G(b_1) + D)}{(v_0^* + a_0) + \left[\dfrac{(1-\beta)v_0^* + a_0}{1 + \left[(1-\beta)v_0^* + a_0\right] K} + a_1\right]\left\{1 + \left[(1-\beta)v_0^* + a_0\right] K\right\}} \\
=& \frac{-(b_1 - b_0) + \left[(1-\beta)v_0^* + a_0\right](G(b_1) + D)}{(v_0^* + a_0) + \left[(1-\beta)v_0^* + a_0\right] + a_1\left\{1 + \left[(1-\beta)v_0^* + a_0\right] K\right\}} \\
=& \frac{-(b_1 - b_0) + \left[(1-\beta)v_0^* + a_0\right](G(b_1) + D)}{(v_0^* + a_0 + a_1) + \left[(1-\beta)v_0^* + a_0\right](1 + a_1 K)} = \frac{M}{N},
\end{aligned}
\tag{A.98}
$$

where

$$
M = M(\beta) = -(b_1 - b_0) + \left[(1-\beta)v_0^* + a_0\right](G(b_1) + D) \tag{A.99}
$$

and

$$N = N(\beta) = \left(v_0^* + a_0 + a_1\right) + \left[(1-\beta)v_0^* + a_0\right](1 + a_1 K). \tag{A.100}$$

It is easy to see that M and N are continuously differentiable with respect to β with

$$M' = \left[-v_0^* + (1-\beta)v_0^{*\prime}\right](G(b_1) + D), \tag{A.101}$$
$$N' = v_0^{*\prime} + \left[-v_0^* + (1-\beta)v_0^{*\prime}\right](1 + a_1 K). \tag{A.102}$$

Moreover, $N > 0$, so q_1^* is continuously differentiable with respect to β and

$$q_1^{*\prime} = \frac{M'N - MN'}{N^2}. \tag{A.103}$$

Thus, to find the value of $q_1^{*\prime}$ it suffices to estimate the value of the numerator of (A.103).

$$\begin{aligned}
&M'N - MN' = \\
&= \left[-v_0^* + (1-\beta)v_0^{*\prime}\right](G(b_1) + D)\left\{\left(v_0^* + a_0 + a_1\right) + \left[(1-\beta)v_0^* + a_0\right](1 + a_1 K)\right\} \\
&\quad - \left\{-(b_1 - b_0) + \left[(1-\beta)v_0^* + a_0\right](G(b_1) + D)\right\}\left\{v_0^{*\prime} + \left[-v_0^* + (1-\beta)v_0^{*\prime}\right](1 + a_1 K)\right\} \\
&= \left(v_0^* + a_0 + a_1\right)\left[-v_0^* + (1-\beta)v_0^{*\prime}\right](G(b_1) + D) \\
&\quad + \left[(1-\beta)v_0^* + a_0\right]\left[-v_0^* + (1-\beta)v_0^{*\prime}\right](1 + a_1 K)(G(b_1) + D) \\
&\quad + \left\{v_0^{*\prime} + \left[-v_0^* + (1-\beta)v_0^{*\prime}\right](1 + a_1 K)\right\}(b_1 - b_0) \\
&\quad - \left\{v_0^{*\prime} + \left[-v_0^* + (1-\beta)v_0^{*\prime}\right](1 + a_1 K)\right\}\left[(1-\beta)v_0^* + a_0\right](G(b_1) + D) \\
&= \left(v_0^* + a_0\right)\left[-v_0^* + (1-\beta)v_0^{*\prime}\right](G(b_1) + D) \\
&\quad + a_1\left[-v_0^* + (1-\beta)v_0^{*\prime}\right](G(b_1) + D) \\
&\quad + \left[(1-\beta)v_0^* + a_0\right]\left[-v_0^* + (1-\beta)v_0^{*\prime}\right](1 + a_1 K)(G(b_1) + D) \\
&\quad + \left\{v_0^{*\prime} + \left[-v_0^* + (1-\beta)v_0^{*\prime}\right](1 + a_1 K)\right\}(b_1 - b_0) \\
&\quad - v_0^{*\prime}\left[(1-\beta)v_0^* + a_0\right](G(b_1) + D) \\
&\quad - \left[(1-\beta)v_0^* + a_0\right]\left[-v_0^* + (1-\beta)v_0^{*\prime}\right](1 + a_1 K)(G(b_1) + D) \\
&= \left(v_0^* + a_0\right)\left[-v_0^* + (1-\beta)v_0^{*\prime}\right](G(b_1) + D) \\
&\quad + a_1\left[-v_0^* + (1-\beta)v_0^{*\prime}\right](G(b_1) + D) \\
&\quad + \left\{v_0^{*\prime} + \left[-v_0^* + (1-\beta)v_0^{*\prime}\right](1 + a_1 K)\right\}(b_1 - b_0) \\
&\quad - v_0^{*\prime}\left[(1-\beta)v_0^* + a_0\right](G(b_1) + D) \\
&= \left\{\left(v_0^* + a_0\right)\left[-v_0^* + (1-\beta)v_0^{*\prime}\right] - v_0^{*\prime}\left[(1-\beta)v_0^* + a_0\right]\right\}(G(b_1) + D) \\
&\quad + a_1\left[-v_0^* + (1-\beta)v_0^{*\prime}\right](G(b_1) + D) \\
&\quad + \left\{v_0^{*\prime} + \left[-v_0^* + (1-\beta)v_0^{*\prime}\right](1 + a_1 K)\right\}(b_1 - b_0) \\
&= \left[\left(v_0^* + a_0\right)\left(-v_0^* - \beta v_0^{*\prime}\right) + \beta v_0^* v_0^{*\prime}\right](G(b_1) + D) \\
&\quad + a_1\left[-v_0^* + (1-\beta)v_0^{*\prime}\right](G(b_1) + D) \\
&\quad + \left\{v_0^{*\prime} + \left[-v_0^* + (1-\beta)v_0^{*\prime}\right](1 + a_1 K)\right\}(b_1 - b_0).
\end{aligned} \tag{A.104}$$

Given the values of a_0, a_1, b_0, b_1, β, v_0^*, $v_0^{*\prime}$, $G(p)$, D and Eq. (A.68), it is clear that (A.104) is non-positive. Moreover,

$$\begin{aligned} M'N - MN' =& \left[\left(v_0^* + a_0\right)\left(-v_0^* - \beta v_0^{*\prime}\right) + \beta v_0^* v_0^{*\prime}\right](G(b_1) + D) \\ &+ a_1\left[-v_0^* + (1-\beta)v_0^{*\prime}\right](G(b_1) + D) \\ &+ \left\{v_0^{*\prime} + \left[-v_0^* + (1-\beta)v_0^{*\prime}\right](1 + a_1 K)\right\}(b_1 - b_0) \\ \le& a_1\left[-v_0^* + (1-\beta)v_0^{*\prime}\right](G(b_1) + D) < 0. \end{aligned} \tag{A.105}$$

Thus,

$$M'N - MN' < 0, \tag{A.106}$$

which proves that $q_1^{*\prime} < 0$, so $q_1^*(\beta)$ is continuously differentiable and strictly decreasing with respect to β, $\beta \in (0, 1]$.

Finally, since

$$q_0^*(\beta) + q_1^*(\beta) = G^*(\beta) + D, \tag{A.107}$$

then,

$$q_0^*(\beta) = -q_1^*(\beta) + G^*(\beta) + D. \tag{A.108}$$

And since $G^*(\beta)$ is continuously differentiable and strictly increasing with respect to β, the function $q_1^*(\beta)$ is continuously differentiable and strictly decreasing with respect to β. Because D is constant, we have that $q_0^*(\beta)$ is continuously differentiable and strictly increasing with respect to β, $\beta \in (0, 1]$ The proof of the theorem is complete ■

Theorem 2.4 *For the affine demand function $G(p)$ described in (2.22), the price $p^c(\beta)$ and the supply values $q_i^c(\beta)$, $i = 0, 1$, from the Cournot-Nash equilibrium, are continuously differentiable with respect to $\beta \in (0, 1]$. Moreover, $p^c(\beta)$ and $q_1^c(\beta)$ strictly decrease, whereas $q_0^c(\beta)$ strictly increase.*

Proof Let's consider the exterior equilibrium (p^c, q_0^c, q_1^c), i.e., such a vector that the following equalities hold:

$$q_0^c + q_1^c = G(p^c) + D, \tag{A.109}$$

$$q_0^c = \frac{p^c - b_0}{(1-\beta)\dfrac{1}{K} + a_0} + \frac{\beta\dfrac{1}{K}}{(1-\beta)\dfrac{1}{K} + a_0}\left(\frac{p^c - b_1}{\dfrac{1}{K} + a_1}\right), \tag{A.110}$$

$$q_1^c = \frac{p^c - b_1}{\dfrac{1}{K} + a_1}, \tag{A.111}$$

where

$$G(p^c) = -Kp^c + T. \tag{A.112}$$

From Eq. (A.109) one has

$$q_0^c + q_1^c - G(p^c) - D = 0. \tag{A.113}$$

By substituting (A.110), (A.111) and (A.112) in (A.113), similarly to (A.4), we have

$$\begin{aligned}
q_0^c + q_1^c - G(p^c) - D =& \frac{p^c - b_0}{(1-\beta)\frac{1}{K} + a_0} + \frac{\beta\frac{1}{K}}{(1-\beta)\frac{1}{K} + a_0}\left(\frac{p^c - b_1}{\frac{1}{K} + a_1}\right) \\
&+ \frac{p^c - b_1}{\frac{1}{K} + a_1} + Kp^c - T - D \\
=& p^c\left[\frac{1}{(1-\beta)\frac{1}{K} + a_0} + \frac{\frac{1}{K} + a_0}{(1-\beta)\frac{1}{K} + a_0}\left(\frac{1}{\frac{1}{K} + a_1}\right)\right] \\
&- \left[\frac{b_0}{(1-\beta)\frac{1}{K} + a_0} + \frac{\frac{1}{K} + a_0}{(1-\beta)\frac{1}{K} + a_0}\left(\frac{b_1}{\frac{1}{K} + a_1}\right)\right] \\
&+ Kp^c - T - D = 0.
\end{aligned} \tag{A.114}$$

Solving (A.114) for p^c, similarly to (A.96), we get the equation

$$\begin{aligned}
p^c =& \frac{\frac{b_0}{(1-\beta)\frac{1}{K} + a_0} + \frac{\frac{1}{K} + a_0}{(1-\beta)\frac{1}{K} + a_0}\left(\frac{b_1}{\frac{1}{K} + a_1}\right) + T + D}{\frac{1}{(1-\beta)\frac{1}{K} + a_0} + \frac{\frac{1}{K} + a_0}{(1-\beta)\frac{1}{K} + a_0}\left(\frac{1}{\frac{1}{K} + a_1}\right) + K} \\
=& \frac{\left(\frac{1}{K} + a_0\right)b_1 + \left(\frac{1}{K} + a_1\right)b_0 + \left[(1-\beta)\frac{1}{K} + a_0\right]\left(\frac{1}{K} + a_1\right)(T + D)}{\left(\frac{1}{K} + a_0\right) + \left(\frac{1}{K} + a_1\right) + \left[(1-\beta)\frac{1}{K} + a_0\right]\left(\frac{1}{K} + a_1\right)K} \\
=& \frac{X}{Y},
\end{aligned} \tag{A.115}$$

where

$$X(\beta) = \left(\frac{1}{K} + a_0\right) b_1 + \left(\frac{1}{K} + a_1\right) b_0 + \left[(1-\beta)\frac{1}{K} + a_0\right]\left(\frac{1}{K} + a_1\right)(T + D) \tag{A.116}$$

and

$$Y(\beta) = \left(\frac{1}{K} + a_0\right) + \left(\frac{1}{K} + a_1\right) + \left[(1-\beta)\frac{1}{K} + a_0\right]\left(\frac{1}{K} + a_1\right) K. \tag{A.117}$$

It's easy to see that X and Y are continuously differentiable with respect to β with

$$X' = -\frac{1}{K}\left(\frac{1}{K} + a_1\right)(T + D), \tag{A.118}$$

$$Y' = -\left(\frac{1}{K} + a_1\right). \tag{A.119}$$

Moreover, $Y > 0$, whence p^c is continuously differentiable with respect to β with

$$p^{c'} = \frac{X'Y - XY'}{Y^2}. \tag{A.120}$$

To compute the value of $p^{c'}$ it is sufficient to calculate the value of the numerator of (A.120):

$$\begin{aligned}
&X'Y - XY' = \\
&= -\tfrac{1}{K}\left(\tfrac{1}{K} + a_1\right)(T + D)\left\{\left(\tfrac{1}{K} + a_0\right) + \left(\tfrac{1}{K} + a_1\right) + \left[(1-\beta)\tfrac{1}{K} + a_0\right]\left(\tfrac{1}{K} + a_1\right) K\right\} \\
&\quad - \left\{\left(\tfrac{1}{K} + a_0\right) b_1 + \left(\tfrac{1}{K} + a_1\right) b_0 + \left[(1-\beta)\tfrac{1}{K} + a_0\right]\left(\tfrac{1}{K} + a_1\right)(T + D)\right\}\left[-\left(\tfrac{1}{K} + a_1\right)\right] \\
&= -\tfrac{1}{K}\left(\tfrac{1}{K} + a_1\right)(T + D)\left\{\left(\tfrac{1}{K} + a_0\right) + \left(\tfrac{1}{K} + a_1\right) + \left[(1-\beta)\tfrac{1}{K} + a_0\right]\left(\tfrac{1}{K} + a_1\right) K\right\} \\
&\quad + \left(\tfrac{1}{K} + a_1\right)\left\{\left(\tfrac{1}{K} + a_0\right) b_1 + \left(\tfrac{1}{K} + a_1\right) b_0 + \left[(1-\beta)\tfrac{1}{K} + a_0\right]\left(\tfrac{1}{K} + a_1\right)(T + D)\right\} \\
&= -\tfrac{1}{K}\left(\tfrac{1}{K} + a_0\right)\left(\tfrac{1}{K} + a_1\right)(T + D) \\
&\quad - \tfrac{1}{K}\left(\tfrac{1}{K} + a_1\right)^2 (T + D) \\
&\quad - \left[(1-\beta)\tfrac{1}{K} + a_0\right]\left(\tfrac{1}{K} + a_1\right)^2 (T + D) \\
&\quad + \left(\tfrac{1}{K} + a_0\right)\left(\tfrac{1}{K} + a_1\right) b_1 \\
&\quad + \left(\tfrac{1}{K} + a_1\right)^2 b_0 \\
&\quad + \left[(1-\beta)\tfrac{1}{K} + a_0\right]\left(\tfrac{1}{K} + a_1\right)^2 (T + D) \\
&= \left(\tfrac{1}{K} + a_0\right)\left(\tfrac{1}{K} + a_1\right) b_1 - \tfrac{1}{K}\left(\tfrac{1}{K} + a_0\right)\left(\tfrac{1}{K} + a_1\right)(T + D) \\
&\quad + \left(\tfrac{1}{K} + a_1\right)^2 b_0 - \tfrac{1}{K}\left(\tfrac{1}{K} + a_1\right)^2 (T + D) \\
&= -\tfrac{1}{K}\left(\tfrac{1}{K} + a_0\right)\left(\tfrac{1}{K} + a_1\right)(-Kb_1 + T + D) \\
&\quad - \tfrac{1}{K}\left(\tfrac{1}{K} + a_1\right)^2 (-Kb_0 + T + D) \\
&= -\tfrac{1}{K}\left(\tfrac{1}{K} + a_0\right)\left(\tfrac{1}{K} + a_1\right)(G(b_1) + D) \\
&\quad - \tfrac{1}{K}\left(\tfrac{1}{K} + a_1\right)^2 (G(b_0) + D).
\end{aligned} \tag{A.121}$$

Given the values of $a_0, a_1, K, G(p)$ and D, it is clear that (A.121) is non-positive. Moreover,

$$\begin{aligned} X'Y - XY' = &-\frac{1}{K}\left(\frac{1}{K} + a_0\right)\left(\frac{1}{K} + a_1\right)(G(b_1) + D) \\ &-\frac{1}{K}\left(\frac{1}{K} + a_1\right)^2 (G(b_0) + D) \\ \leq &-\frac{1}{K}\left(\frac{1}{K} + a_0\right)\left(\frac{1}{K} + a_1\right)(G(b_1) + D) < 0. \end{aligned} \tag{A.122}$$

Then,

$$X'Y - XY' < 0, \tag{A.123}$$

which proves that $p^{c\prime} < 0$, so $p^c(\beta)$ is continuously differentiable and strictly decreasing with respect to β, $\beta \in (0, 1]$.

Since

$$q_1^c(\beta) = \frac{p^c(\beta) - b_1}{\dfrac{1}{K} + a_1}, \tag{A.124}$$

and $p^c(\beta)$ is continuously differentiable and strictly decreasing with respect to β, and a_1, b_1 and K re positive constants, then, $q_1^c(\beta)$ is continuously differentiable and strictly decreasing with respect to β, $\beta \in (0, 1]$.

Finally, since

$$q_0^c(\beta) + q_1^c(\beta) = G(p^c(\beta)) + D = -Kp^c(\beta) + T + D, \tag{A.125}$$

then,

$$q_0^c(\beta) = -q_1^c(\beta) - Kp^c(\beta) + T + D. \tag{A.126}$$

And as $q_1^c(\beta)$ is continuously differentiable and strictly decreasing with respect to β, the function $p^c(\beta)$ also has the same property, and K, T and D are non-negative constants, then, $q_0^*(\beta)$ is continuously differentiable and strictly increasing with respect to β, $\beta \in (0, 1]$. The proof of the theorem is complete ■

Theorem 2.5 *For the affine demand function $G(p)$ described in (2.22), the price $p^t(\beta)$ and the output volumes $q_i^t(\beta)$, $i = 0, 1$, related to the perfect competition equilibrium, are invariant for all $\beta \in (0, 1]$ and are described by the clear-cut expressions:*

$$p^t = \frac{a_0 b_1 + a_1 b_0 + a_0 a_1 (T + D)}{a_0 + a_1 + a_0 a_1 K}, \tag{A.127}$$

$$q_0^t = \frac{a_1 (G(b_0) + D) + (b_1 - b_0)}{a_0 + a_1 + a_0 a_1 K}, \tag{A.128}$$

$$q_1^t = \frac{a_0 (G(b_1) + D) - (b_1 - b_0)}{a_0 + a_1 + a_0 a_1 K}. \tag{A.129}$$

Proof Let us consider the exterior equilibrium (p^t, q_0^t, q_1^t), i.e., such a vector that the following equalities hold:

$$q_0^t + q_1^t = G(p^t) + D, \tag{A.130}$$

$$q_0^t = \frac{p^t - b_0}{a_0}, \tag{A.131}$$

$$q_1^t = \frac{p^t - b_1}{a_1}, \tag{A.132}$$

where

$$G(p^t) = -Kp^t + T. \tag{A.133}$$

From (A.109) one gets that

$$q_0^t + q_1^t - G(p^t) - D = 0. \tag{A.134}$$

Next, by plugging (A.131), (A.132) and (A.133) in (A.134), we deduce that

$$\begin{aligned} q_0^t + q_1^t - G(p^t) - D =& \frac{p^t - b_0}{a_0} + \frac{p^t - b_1}{a_1} + Kp^t - T - D \\ =& p^t \left(\frac{1}{a_0} + \frac{1}{a_1} \right) - \left(\frac{b_1}{a_1} + \frac{b_0}{a_0} \right) + Kp^c - T - D = 0. \end{aligned} \tag{A.135}$$

By solving Eq. (A.135) for p^t, we obtain the equality

$$p^t = \frac{\frac{b_1}{a_1} + \frac{b_0}{a_0} + T + D}{\frac{1}{a_0} + \frac{1}{a_1} + K} = \frac{a_0 b_1 + a_1 b_0 + a_0 a_1 (T + D)}{a_0 + a_1 + a_0 a_1 K}, \tag{A.136}$$

showing that the function $p^t(\beta)$ is constant for all $\beta \in (0, 1]$.

Moreover, since

$$
\begin{aligned}
q_0^t &= \frac{p^t - b_0}{a_0} = \frac{\dfrac{a_0 b_1 + a_1 b_0 + a_0 a_1 (T + D)}{a_0 + a_1 + a_0 a_1 K} - b_0}{a_0} \\
&= \frac{a_0 b_1 + a_1 b_0 + a_0 a_1 (T + D) - (a_0 + a_1 + a_0 a_1 K)\, b_0}{a_0\,(a_0 + a_1 + a_0 a_1 K)} \\
&= \frac{a_0\,(b_1 - b_0) + a_0 a_1\,(-K b_0 + T + D)}{a_0\,(a_0 + a_1 + a_0 a_1 K)} \\
&= \frac{a_1\,(G(b_0) + D) + (b_1 - b_0)}{a_0 + a_1 + a_0 a_1 K},
\end{aligned}
\tag{A.137}
$$

and

$$
\begin{aligned}
q_1^t &= \frac{p^t - b_1}{a_1} = \frac{\dfrac{a_0 b_1 + a_1 b_0 + a_0 a_1 (T + D)}{a_0 + a_1 + a_0 a_1 K} - b_1}{a_1} \\
&= \frac{a_0 b_1 + a_1 b_0 + a_0 a_1 (T + D) - (a_0 + a_1 + a_0 a_1 K)\, b_1}{a_1\,(a_0 + a_1 + a_0 a_1 K)} \\
&= \frac{-a_1\,(b_1 - b_0) + a_0 a_1\,(-K b_1 + T + D)}{a_1\,(a_0 + a_1 + a_0 a_1 K)} \\
&= \frac{a_0\,(G(b_1) + D) - (b_1 - b_0)}{a_0 + a_1 + a_0 a_1 K},
\end{aligned}
\tag{A.138}
$$

the functions $q_0^t(\beta)$ and $q_1^t(\beta)$ are constant for all $\beta \in (0, 1]$, too. The proof of the theorem is complete ■

Theorem 2.6 *For the affine demand function $G(p)$ from (2.22), the price functions in the CCVE, $p^*(\beta)$, the Cournot-Nash equilibrium, $p^c(\beta)$, and the perfect competition equilibrium, p^t, satisfy the following inequalities:*

$$
p^t < \lim_{\beta \to 0} p^*(\beta), \tag{A.139}
$$

and

$$
p^*(\beta) < p^c(\beta), \quad \forall \beta \in (0, 1]. \tag{A.140}
$$

Proof First, we prove inequality (A.139):

$$
p^t < \lim_{\beta \to 0} p^*(\beta).
$$

Introduce the following notation:

$$
\begin{aligned}
\widehat{v_0^*} &= \lim_{\beta \to 0} v_0^*(\beta) \\
&= \frac{2(a_0 + a_1 + a_0 a_1 K)}{(2a_0 K + a_0 a_1 K^2) + \sqrt{(2a_0 K + a_0 a_1 K^2)^2 + 4(2K + a_1 K^2)(a_0 + a_1 + a_0 a_1 K)}} > 0,
\end{aligned}
\tag{A.141}
$$

$$\widehat{\nu_1^*} = \lim_{\beta \to 0} \nu_1^*(\beta) = \lim_{\beta \to 0} \frac{(1-\beta)\nu_0^* + a_0}{1 + \left[(1-\beta)\nu_0^* + a_0\right] K} = \frac{\widehat{\nu_0^*} + a_0}{1 + \left(\widehat{\nu_0^*} + a_0\right) K} > 0. \tag{A.142}$$

Therefore,

$$\begin{aligned} \lim_{\beta \to 0} p^*(\beta) &= \lim_{\beta \to 0} \frac{(\nu_0^* + a_0)\, b_1 + (\nu_1^* + a_1)\, b_0 + \left[(1-\beta)\nu_0^* + a_0\right](\nu_1^* + a_1)(T+D)}{(\nu_0^* + a_0) + (\nu_1^* + a_1) + \left[(1-\beta)\nu_0^* + a_0\right](\nu_1^* + a_1) K} \\ &= \frac{\left(\widehat{\nu_0^*} + a_0\right) b_1 + \left(\widehat{\nu_1^*} + a_1\right) b_0 + \left(\widehat{\nu_0^*} + a_0\right)\left(\widehat{\nu_1^*} + a_1\right)(T+D)}{\left(\widehat{\nu_0^*} + a_0\right) + \left(\widehat{\nu_1^*} + a_1\right) + \left(\widehat{\nu_0^*} + a_0\right)\left(\widehat{\nu_1^*} + a_1\right) K}. \end{aligned} \tag{A.143}$$

Now, we compute the difference

$$\begin{aligned} &\lim_{\beta \to 0} p^*(\beta) - p^t = \\ &= \frac{\left(\widehat{\nu_0^*} + a_0\right) b_1 + \left(\widehat{\nu_1^*} + a_1\right) b_0 + \left(\widehat{\nu_0^*} + a_0\right)\left(\widehat{\nu_1^*} + a_1\right)(T+D)}{\left(\widehat{\nu_0^*} + a_0\right) + \left(\widehat{\nu_1^*} + a_1\right) + \left(\widehat{\nu_0^*} + a_0\right)\left(\widehat{\nu_1^*} + a_1\right) K} \\ &\quad - \frac{a_0 b_1 + a_1 b_0 + a_0 a_1 (T+D)}{a_0 + a_1 + a_0 a_1 K} \\ &= \frac{\left[\left(\widehat{\nu_0^*} + a_0\right) b_1 + \left(\widehat{\nu_1^*} + a_1\right) b_0 + \left(\widehat{\nu_0^*} + a_0\right)\left(\widehat{\nu_1^*} + a_1\right)(T+D)\right](a_0 + a_1 + a_0 a_1 K)}{\left[\left(\widehat{\nu_0^*} + a_0\right) + \left(\widehat{\nu_1^*} + a_1\right) + \left(\widehat{\nu_0^*} + a_0\right)\left(\widehat{\nu_1^*} + a_1\right) K\right](a_0 + a_1 + a_0 a_1 K)} \\ &\quad - \frac{\left[\left(\widehat{\nu_0^*} + a_0\right) + \left(\widehat{\nu_1^*} + a_1\right) + \left(\widehat{\nu_0^*} + a_0\right)\left(\widehat{\nu_1^*} + a_1\right) K\right][a_0 b_1 + a_1 b_0 + a_0 a_1 (T+D)]}{\left[\left(\widehat{\nu_0^*} + a_0\right) + \left(\widehat{\nu_1^*} + a_1\right) + \left(\widehat{\nu_0^*} + a_0\right)\left(\widehat{\nu_1^*} + a_1\right) K\right](a_0 + a_1 + a_0 a_1 K)} \\ &= \frac{R_1}{R_2}, \end{aligned} \tag{A.144}$$

where

$$\begin{aligned} R_1 = &\left[\left(\widehat{\nu_0^*} + a_0\right) b_1 + \left(\widehat{\nu_1^*} + a_1\right) b_0 + \left(\widehat{\nu_0^*} + a_0\right)\left(\widehat{\nu_1^*} + a_1\right)(T+D)\right](a_0 + a_1 + a_0 a_1 K) \\ &- \left[\left(\widehat{\nu_0^*} + a_0\right) + \left(\widehat{\nu_1^*} + a_1\right) + \left(\widehat{\nu_0^*} + a_0\right)\left(\widehat{\nu_1^*} + a_1\right) K\right][a_0 b_1 + a_1 b_0 + a_0 a_1 (T+D)] \end{aligned} \tag{A.145}$$

and

$$R_2 = \left[\left(\widehat{\nu_0^*} + a_0\right) + \left(\widehat{\nu_1^*} + a_1\right) + \left(\widehat{\nu_0^*} + a_0\right)\left(\widehat{\nu_1^*} + a_1\right) K\right](a_0 + a_1 + a_0 a_1 K). \tag{A.146}$$

Given the values of a_0, a_1, $\widehat{\nu_0^*}$, $\widehat{\nu_1^*}$, and K, it is easy to see that $R_2 > 0$. Hence, to calculate the value of (A.144), it is enough to estimate the value of (A.145). That is,

$$
\begin{aligned}
R_1 =& \left[\left(\widehat{v_0^*}+a_0\right)b_1+\left(\widehat{v_1^*}+a_1\right)b_0+\left(\widehat{v_0^*}+a_0\right)\left(\widehat{v_1^*}+a_1\right)(T+D)\right](a_0+a_1+a_0a_1K)\\
&-\left[\left(\widehat{v_0^*}+a_0\right)+\left(\widehat{v_1^*}+a_1\right)+\left(\widehat{v_0^*}+a_0\right)\left(\widehat{v_1^*}+a_1\right)K\right][a_0b_1+a_1b_0+a_0a_1(T+D)]\\
=&(a_0+a_1+a_0a_1K)\left(\widehat{v_0^*}+a_0\right)b_1\\
&+(a_0+a_1+a_0a_1K)\left(\widehat{v_1^*}+a_1\right)b_0\\
&+(a_0+a_1+a_0a_1K)\left(\widehat{v_0^*}+a_0\right)\left(\widehat{v_1^*}+a_1\right)(T+D)\\
&-a_0\left[\left(\widehat{v_0^*}+a_0\right)+\left(\widehat{v_1^*}+a_1\right)+\left(\widehat{v_0^*}+a_0\right)\left(\widehat{v_1^*}+a_1\right)K\right]b_1\\
&-a_1\left[\left(\widehat{v_0^*}+a_0\right)+\left(\widehat{v_1^*}+a_1\right)+\left(\widehat{v_0^*}+a_0\right)\left(\widehat{v_1^*}+a_1\right)K\right]b_0\\
&-a_0a_1\left[\left(\widehat{v_0^*}+a_0\right)+\left(\widehat{v_1^*}+a_1\right)+\left(\widehat{v_0^*}+a_0\right)\left(\widehat{v_1^*}+a_1\right)K\right](T+D)\\
=&\left\{(a_1+a_0a_1K)\left(\widehat{v_0^*}+a_0\right)-a_0\left[\left(\widehat{v_1^*}+a_1\right)+\left(\widehat{v_0^*}+a_0\right)\left(\widehat{v_1^*}+a_1\right)K\right]\right\}b_1\\
&+\left\{(a_0+a_0a_1K)\left(\widehat{v_1^*}+a_1\right)-a_1\left[\left(\widehat{v_0^*}+a_0\right)+\left(\widehat{v_0^*}+a_0\right)\left(\widehat{v_1^*}+a_1\right)K\right]\right\}b_0\\
&+\left\{(a_0+a_1)\left(\widehat{v_0^*}+a_0\right)\left(\widehat{v_1^*}+a_1\right)-a_0a_1\left[\left(\widehat{v_0^*}+a_0\right)+\left(\widehat{v_1^*}+a_1\right)\right]\right\}(T+D)\\
=&\left[a_1\widehat{v_0^*}-a_0\widehat{v_1^*}-a_0\widehat{v_1^*}\left(\widehat{v_0^*}+a_0\right)K\right]b_1\\
&+\left[a_0\widehat{v_1^*}-a_1\widehat{v_0^*}-a_1\widehat{v_0^*}\left(\widehat{v_1^*}+a_1\right)K\right]b_0\\
&+\left[a_0\widehat{v_1^*}\left(\widehat{v_0^*}+a_0\right)+a_1\widehat{v_0^*}\left(\widehat{v_1^*}+a_1\right)\right](T+D)\\
=&-a_0\widehat{v_1^*}\left(\widehat{v_0^*}+a_0\right)Kb_1+a_0\widehat{v_1^*}\left(\widehat{v_0^*}+a_0\right)(T+D)\\
&-a_1\widehat{v_0^*}\left(\widehat{v_1^*}+a_1\right)Kb_0+a_1\widehat{v_0^*}\left(\widehat{v_1^*}+a_1\right)(T+D)\\
&+\left(a_1\widehat{v_0^*}-a_0\widehat{v_1^*}\right)b_1+\left(a_0\widehat{v_1^*}-a_1\widehat{v_0^*}\right)b_0\\
=&a_0\widehat{v_1^*}\left(\widehat{v_0^*}+a_0\right)(-Kb_1+T+D)\\
&+a_1\widehat{v_0^*}\left(\widehat{v_1^*}+a_1\right)(-Kb_0+T+D)\\
&+\left(a_1\widehat{v_0^*}-a_0\widehat{v_1^*}\right)(b_1-b_0)\\
=&a_0\widehat{v_1^*}\left(\widehat{v_0^*}+a_0\right)(G(b_1)+D)-a_0\widehat{v_1^*}(b_1-b_0)\\
&+a_1\widehat{v_0^*}\left(\widehat{v_1^*}+a_1\right)(G(b_0)+D)+a_1\widehat{v_0^*}(b_1-b_0)\\
=&a_0\widehat{v_1^*}\left[\left(\widehat{v_0^*}+a_0\right)(G(b_1)+D)-(b_1-b_0)\right]\\
&+a_1\widehat{v_0^*}\left[\left(\widehat{v_1^*}+a_1\right)(G(b_0)+D)+(b_1-b_0)\right].
\end{aligned}
\tag{A.147}
$$

Given the values of a_0, a_1, b_0, b_1, $\widehat{v_0^*}$, $\widehat{v_1^*}$, $G(p)$, D and assumption **A**3, it is trivial that (A.147) is nonnegative. Moreover,

$$\begin{aligned}R_1 =&a_0\widehat{v_1^*}\left[\left(\widehat{v_0^*}+a_0\right)\left(G(b_1)+D\right)-(b_1-b_0)\right]\\&+a_1\widehat{v_0^*}\left[\left(\widehat{v_1^*}+a_1\right)\left(G(b_0)+D\right)+(b_1-b_0)\right]\\\geq&a_1\widehat{v_0^*}\left[\left(\widehat{v_1^*}+a_1\right)\left(G(b_0)+D\right)+(b_1-b_0)\right]\geq a_1\widehat{v_0^*}\left(\widehat{v_1^*}+a_1\right)\left(G(b_0)+D\right)\\\geq&a_1^2\widehat{v_0^*}\left(G(b_0)+D\right)\geq a_1^2\widehat{v_0^*}G(b_0)>0.\end{aligned} \tag{A.148}$$

And since $R_1 > 0$, by (A.148), then,

$$\lim_{\beta\to 0} p^*(\beta) - p^t > 0, \tag{A.149}$$

which proves inequality (A.139).

Now, we establish inequality (A.140):

$$p^*(\beta) < p^c(\beta) \text{ para todo } \beta \in (0, 1].$$

In order to do that, we introduce the following notation:

$$v_i^* = v_i^*(\beta), i = 0, 1.$$

From Eqs. (A.42) and (A.43) it's easy to see that the following inequality hold for all $\beta \in (0, 1]$:

$$v_i^* < \frac{1}{K}, i = 0, 1. \tag{A.150}$$

Now, we compute the difference

$$\begin{aligned}&\left(p^c - p^*\right)(\beta) =\\&=\frac{\left(\frac{1}{K}+a_0\right)b_1+\left(\frac{1}{K}+a_1\right)b_0+\left[(1-\beta)\frac{1}{K}+a_0\right]\left(\frac{1}{K}+a_1\right)(T+D)}{\left(\frac{1}{K}+a_0\right)+\left(\frac{1}{K}+a_1\right)+\left[(1-\beta)\frac{1}{K}+a_0\right]\left(\frac{1}{K}+a_1\right)K}\\&-\frac{\left(v_0^*+a_0\right)b_1+\left(v_1^*+a_1\right)b_0+\left[(1-\beta)v_0^*+a_0\right]\left(v_1^*+a_1\right)(T+D)}{\left(v_0^*+a_0\right)+\left(v_1^*+a_1\right)+\left[(1-\beta)v_0^*+a_0\right]\left(v_1^*+a_1\right)K}\\&=\left\langle\left\{\left(\frac{1}{K}+a_0\right)b_1+\left(\frac{1}{K}+a_1\right)b_0+\left[(1-\beta)\frac{1}{K}+a_0\right]\left(\frac{1}{K}+a_1\right)(T+D)\right\}\times\right.\\&\left\{\left(v_0^*+a_0\right)+\left(v_1^*+a_1\right)+\left[(1-\beta)v_0^*+a_0\right]\left(v_1^*+a_1\right)K\right\}\\&-\left\{\left(\frac{1}{K}+a_0\right)+\left(\frac{1}{K}+a_1\right)+\left[(1-\beta)\frac{1}{K}+a_0\right]\left(\frac{1}{K}+a_1\right)K\right\}\times\\&\left.\left\{\left(v_0^*+a_0\right)b_1+\left(v_1^*+a_1\right)b_0+\left[(1-\beta)v_0^*+a_0\right]\left(v_1^*+a_1\right)(T+D)\right\}\right\rangle\Big/\\&\left\langle\left\{\left(\frac{1}{K}+a_0\right)+\left(\frac{1}{K}+a_1\right)+\left[(1-\beta)\frac{1}{K}+a_0\right]\left(\frac{1}{K}+a_1\right)K\right\}\times\right.\\&\left.\left\{\left(v_0^*+a_0\right)+\left(v_1^*+a_1\right)+\left[(1-\beta)v_0^*+a_0\right]\left(v_1^*+a_1\right)K\right\}\right\rangle\\&=\frac{S_1}{S_2},\end{aligned} \tag{A.151}$$

where

$$
\begin{aligned}
S_1 = & \left\{\left(\frac{1}{K}+a_0\right)b_1+\left(\frac{1}{K}+a_1\right)b_0+\left[(1-\beta)\frac{1}{K}+a_0\right]\left(\frac{1}{K}+a_1\right)(T+D)\right\}\times \\
& \left\{\left(v_0^*+a_0\right)+\left(v_1^*+a_1\right)+\left[(1-\beta)v_0^*+a_0\right]\left(v_1^*+a_1\right)K\right\} \\
& -\left\{\left(\frac{1}{K}+a_0\right)+\left(\frac{1}{K}+a_1\right)+\left[(1-\beta)\frac{1}{K}+a_0\right]\left(\frac{1}{K}+a_1\right)K\right\}\times \\
& \left\{\left(v_0^*+a_0\right)b_1+\left(v_1^*+a_1\right)b_0+\left[(1-\beta)v_0^*+a_0\right]\left(v_1^*+a_1\right)(T+D)\right\}
\end{aligned} \tag{A.152}
$$

and

$$
\begin{aligned}
S_2 = & \left\{\left(\tfrac{1}{K}+a_0\right)+\left(\tfrac{1}{K}+a_1\right)+\left[(1-\beta)\tfrac{1}{K}+a_0\right]\left(\tfrac{1}{K}+a_1\right)K\right\}\times \\
& \left\{\left(v_0^*+a_0\right)+\left(v_1^*+a_1\right)+\left[(1-\beta)v_0^*+a_0\right]\left(v_1^*+a_1\right)K\right\}.
\end{aligned} \tag{A.153}
$$

For any fixed values of $a_0, a_1, \beta, v_0^*, v_1^*$, and K, it is apparent that $S_2 > 0$. Because of that, in order to find the value of (A.151), it suffices to calculate the value of (A.152). So,

$$
\begin{aligned}
S_1 = & \left\{\left(v_0^*+a_0\right)+\left(v_1^*+a_1\right)+\left[(1-\beta)v_0^*+a_0\right]\left(v_1^*+a_1\right)K\right\}\left(\tfrac{1}{K}+a_0\right)b_1 \\
& +\left\{\left(v_0^*+a_0\right)+\left(v_1^*+a_1\right)+\left[(1-\beta)v_0^*+a_0\right]\left(v_1^*+a_1\right)K\right\}\left(\tfrac{1}{K}+a_1\right)b_0 \\
& +\left\{\left(v_0^*+a_0\right)+\left(v_1^*+a_1\right)+\left[(1-\beta)v_0^*+a_0\right]\left(v_1^*+a_1\right)K\right\}\left[(1-\beta)\tfrac{1}{K}+a_0\right]\left(\tfrac{1}{K}+a_1\right)(T+D) \\
& -\left\{\left(\tfrac{1}{K}+a_0\right)+\left(\tfrac{1}{K}+a_1\right)+\left[(1-\beta)\tfrac{1}{K}+a_0\right]\left(\tfrac{1}{K}+a_1\right)K\right\}\left(v_0^*+a_0\right)b_1 \\
& -\left\{\left(\tfrac{1}{K}+a_0\right)+\left(\tfrac{1}{K}+a_1\right)+\left[(1-\beta)\tfrac{1}{K}+a_0\right]\left(\tfrac{1}{K}+a_1\right)K\right\}\left(v_1^*+a_1\right)b_0 \\
& -\left\{\left(\tfrac{1}{K}+a_0\right)+\left(\tfrac{1}{K}+a_1\right)+\left[(1-\beta)\tfrac{1}{K}+a_0\right]\left(\tfrac{1}{K}+a_1\right)K\right\}\left[(1-\beta)v_0^*+a_0\right]\left(v_1^*+a_1\right)(T+D) \\
= & \left[(1-\beta)\tfrac{1}{K}+a_0\right]\left(v_0^*+a_0\right)\left(\tfrac{1}{K}+a_1\right)(-Kb_1+T+D) \\
& -\left[(1-\beta)v_0^*+a_0\right]\left(v_1^*+a_1\right)\left(\tfrac{1}{K}+a_0\right)(-Kb_1+T+D) \\
& +\left[(1-\beta)\tfrac{1}{K}+a_0\right]\left(v_1^*+a_1\right)\left(\tfrac{1}{K}+a_1\right)(-Kb_0+T+D) \\
& -\left[(1-\beta)v_0^*+a_0\right]\left(v_1^*+a_1\right)\left(\tfrac{1}{K}+a_1\right)(-Kb_0+T+D) \\
& +\left(v_1^*+a_1\right)\left(\tfrac{1}{K}+a_0\right)(b_1-b_0)-\left(v_0^*+a_0\right)\left(\tfrac{1}{K}+a_1\right)(b_1-b_0) \\
= & \left\{\left[(1-\beta)\tfrac{1}{K}+a_0\right]\left(v_0^*+a_0\right)\left(\tfrac{1}{K}+a_1\right)-\left[(1-\beta)v_0^*+a_0\right]\left(v_1^*+a_1\right)\left(\tfrac{1}{K}+a_0\right)\right\}(-Kb_1+T+D) \\
& +\left\{\left[(1-\beta)\tfrac{1}{K}+a_0\right]\left(v_1^*+a_1\right)\left(\tfrac{1}{K}+a_1\right)-\left[(1-\beta)v_0^*+a_0\right]\left(v_1^*+a_1\right)\left(\tfrac{1}{K}+a_1\right)\right\}(-Kb_0+T+D) \\
& +\left[\left(v_1^*+a_1\right)\left(\tfrac{1}{K}+a_0\right)-\left(v_0^*+a_0\right)\left(\tfrac{1}{K}+a_1\right)\right](b_1-b_0) \\
= & \left\{\left[(1-\beta)\tfrac{1}{K}+a_0\right]\left(v_0^*+a_0\right)\left(\tfrac{1}{K}+a_1\right)-\left[(1-\beta)v_0^*+a_0\right]\left(v_1^*+a_1\right)\left(\tfrac{1}{K}+a_0\right)\right\}(G(b_1)+D) \\
& +\left\{\left[(1-\beta)\tfrac{1}{K}+a_0\right]\left(v_1^*+a_1\right)\left(\tfrac{1}{K}+a_1\right)-\left[(1-\beta)v_0^*+a_0\right]\left(v_1^*+a_1\right)\left(\tfrac{1}{K}+a_1\right)\right\}(G(b_0)+D) \\
& +\left[\left(v_1^*+a_1\right)\left(\tfrac{1}{K}+a_0\right)-\left(v_0^*+a_0\right)\left(\tfrac{1}{K}+a_1\right)\right](b_1-b_0) \\
= & X_1(G(b_1)+D)+X_2(G(b_0)+D)+X_3(b_1-b_0),
\end{aligned} \tag{A.154}
$$

where

$$X_1 = \left[(1-\beta)\tfrac{1}{K} + a_0\right](v_0^* + a_0)\left(\tfrac{1}{K} + a_1\right) - \left[(1-\beta)v_0^* + a_0\right](v_1^* + a_1)\left(\tfrac{1}{K} + a_0\right), \tag{A.155}$$

$$X_2 = \left[(1-\beta)\tfrac{1}{K} + a_0\right](v_1^* + a_1)\left(\tfrac{1}{K} + a_1\right) - \left[(1-\beta)v_0^* + a_0\right](v_1^* + a_1)\left(\tfrac{1}{K} + a_1\right) \tag{A.156}$$

and

$$X_3 = (v_1^* + a_1)\left(\tfrac{1}{K} + a_0\right) - (v_0^* + a_0)\left(\tfrac{1}{K} + a_1\right). \tag{A.157}$$

Now, for any given values of a_0, a_1, β, v_0^*, v_1^*, K and (A.150), one finds

$$\begin{aligned} X_2 =& \left[(1-\beta)\tfrac{1}{K} + a_0\right](v_1^* + a_1)\left(\tfrac{1}{K} + a_1\right) - \left[(1-\beta)v_0^* + a_0\right](v_1^* + a_1)\left(\tfrac{1}{K} + a_1\right) \\ =& (1-\beta)\left(\tfrac{1}{K} - v_0^*\right)(v_1^* + a_1)\left(\tfrac{1}{K} + a_1\right) \geq 0 \end{aligned} \tag{A.158}$$

for all $\beta \in (0, 1]$.

Now, we are going to show that $X_1 > 0$ for all $\beta \in (0, 1]$. Plugging (A.42) in X_1, we obtain

$$\begin{aligned} X_1 =& \left[(1-\beta)\tfrac{1}{K} + a_0\right](v_0^* + a_0)\left(\tfrac{1}{K} + a_1\right) \\ & - \left[(1-\beta)v_0^* + a_0\right]\left(\frac{(1-\beta)v_0^* + a_0}{1 + \left[(1-\beta)v_0^* + a_0\right]K} + a_1\right)\left(\tfrac{1}{K} + a_0\right) \\ =& \frac{\left[(1-\beta)\tfrac{1}{K} + a_0\right](v_0^* + a_0)\left(\tfrac{1}{K} + a_1\right)\left\{1 + \left[(1-\beta)v_0^* + a_0\right]K\right\}}{1 + \left[(1-\beta)v_0^* + a_0\right]K} \\ & - \frac{\left[(1-\beta)v_0^* + a_0\right]\left[(1-\beta)v_0^* + a_0 + a_1\left\{1 + \left[(1-\beta)v_0^* + a_0\right]K\right\}\right]\left(\tfrac{1}{K} + a_0\right)}{1 + \left[(1-\beta)v_0^* + a_0\right]K} \\ =& \frac{T_1}{T_2}, \end{aligned} \tag{A.159}$$

where

$$\begin{aligned} T_1 =& \left[(1-\beta)\tfrac{1}{K} + a_0\right](v_0^* + a_0)\left(\tfrac{1}{K} + a_1\right)\left\{1 + \left[(1-\beta)v_0^* + a_0\right]K\right\} \\ & - \left[(1-\beta)v_0^* + a_0\right]\left[(1-\beta)v_0^* + a_0 + a_1\left\{1 + \left[(1-\beta)v_0^* + a_0\right]K\right\}\right]\left(\tfrac{1}{K} + a_0\right) \end{aligned} \tag{A.160}$$

and

$$T_2 = 1 + \left[(1-\beta)v_0^* + a_0\right]K. \tag{A.161}$$

For any fixed values of a_0, β, v_0^*, and K, it is clear that $T_2 > 0$. Therefore, to compute the value of (A.159), we need to calculate the value of T_1.

$$
\begin{aligned}
T_1 =& \left[(1-\beta)\tfrac{1}{K}+a_0\right](v_0^*+a_0)\left(\tfrac{1}{K}+a_1\right)\{1+[(1-\beta)v_0^*+a_0]K\} \\
& -[(1-\beta)v_0^*+a_0]\left[(1-\beta)v_0^*+a_0+a_1\{1+[(1-\beta)v_0^*+a_0]K\}\right]\left(\tfrac{1}{K}+a_0\right) \\
=& \left[(1-\beta)\tfrac{1}{K}+a_0\right]\left(\tfrac{1}{K}+a_1\right)(v_0^*+a_0) \\
& +\left[(1-\beta)^2\tfrac{1}{K}v_0^*+(1-\beta)a_0\left(\tfrac{1}{K}+v_0^*\right)+a_0^2\right]\left(\tfrac{1}{K}+a_1\right)(v_0^*+a_0)K \\
& -a_1[(1-\beta)v_0^*+a_0]\left(\tfrac{1}{K}+a_0\right) \\
& -\left[(1-\beta)^2v_0^{*2}+2(1-\beta)a_0v_0^*+a_0^2\right](1+a_1K)\left(\tfrac{1}{K}+a_0\right) \\
=& (1-\beta)^2v_0^*\left[\left(\tfrac{1}{K}+a_1\right)(v_0^*+a_0)-\left(\tfrac{1}{K}+a_0\right)(1+a_1K)v_0^*\right] \\
& +(1-\beta)\Big[\left(\tfrac{1}{K}+a_1\right)(v_0^*+a_0)\tfrac{1}{K}+\left(\tfrac{1}{K}+a_1\right)(v_0^*+a_0)\left(\tfrac{1}{K}+v_0^*\right)a_0K \\
& \qquad -\left(\tfrac{1}{K}+a_0\right)a_1v_0^*-2\left(\tfrac{1}{K}+a_0\right)(1+a_1K)a_0v_0^*\Big] \\
& +a_0\Big[\left(\tfrac{1}{K}+a_1\right)(v_0^*+a_0)a_0K+\left(\tfrac{1}{K}+a_1\right)(v_0^*+a_0) \\
& \qquad -\left(\tfrac{1}{K}+a_0\right)a_1-\left(\tfrac{1}{K}+a_0\right)(1+a_1K)a_0\Big] \\
=& (1-\beta)^2\left(\tfrac{1}{K}+a_1\right)\left(\tfrac{1}{K}-v_0^*\right)a_0Kv_0^* \\
& +(1-\beta)\Big[\left(\tfrac{1}{K}+a_1\right)(v_0^*+a_0)\tfrac{1}{K}+\left(\tfrac{1}{K}+a_1\right)(v_0^*+a_0)\left(\tfrac{1}{K}+v_0^*\right)a_0K \\
& \qquad -\left(\tfrac{1}{K}+a_0\right)a_1v_0^*-2\left(\tfrac{1}{K}+a_0\right)(1+a_1K)a_0v_0^*\Big] \\
& +\left(\tfrac{1}{K}+a_0\right)a_0\left[\left(\tfrac{1}{K}+a_1\right)Kv_0^*-a_1\right] \\
=& (1-\beta)^2Y_1+(1-\beta)Y_2+\left(\tfrac{1}{K}+a_0\right)a_0Y_3,
\end{aligned}
\tag{A.162}
$$

where

$$Y_1=\left(\tfrac{1}{K}+a_1\right)\left(\tfrac{1}{K}-v_0^*\right)a_0Kv_0^*, \tag{A.163}$$

$$
\begin{aligned}
Y_2 =& \left(\tfrac{1}{K}+a_1\right)(v_0^*+a_0)\tfrac{1}{K}+\left(\tfrac{1}{K}+a_1\right)(v_0^*+a_0)\left(\tfrac{1}{K}+v_0^*\right)a_0K \\
& -\left(\tfrac{1}{K}+a_0\right)a_1v_0^*-2\left(\tfrac{1}{K}+a_0\right)(1+a_1K)a_0v_0^*
\end{aligned}
\tag{A.164}
$$

and

$$Y_3=\left(\tfrac{1}{K}+a_1\right)Kv_0^*-a_1. \tag{A.165}$$

For any fixed values of a_0, a_1, v_0^*, K, and (A.150), one concludes that $Y_1>0$, and $Y_3=Y_3(\beta)$ strictly decreases by β, since $v_0^*=v_0^*(\beta)$ is strictly decreasing with respect to β, and $\left(\tfrac{1}{K}+a_1\right)K>0$. Thus,

$$
\begin{aligned}
Y_3 =&Y_3(\beta) \geq Y_3(1) = \left(\tfrac{1}{K} + a_1\right) K v_0^*(1) - a_1 \\
=&\left(\tfrac{1}{K} + a_1\right) K \frac{a_0 + a_1 + a_0 a_1 K}{1 + 2a_0 K + a_1 K + a_0 a_1 K^2} - a_1 \\
=&\frac{a_0 + a_1 + a_0 a_1 K}{1 + 2a_0 K + a_1 K + a_0 a_1 K^2} (1 + a_1 K) - a_1 \\
=&\frac{(a_0 + a_1 + a_0 a_1 K)(1 + a_1 K) - \left(1 + 2a_0 K + a_1 K + a_0 a_1 K^2\right) a_1}{1 + 2a_0 K + a_1 K + a_0 a_1 K^2} \\
=&\frac{[(1 + a_1 K) a_0 + a_1](1 + a_1 K) - [(1 + a_1 K) + (2 + a_1 K) a_0 K] a_1}{1 + 2a_0 K + a_1 K + a_0 a_1 K^2} \\
=&\frac{(1 + a_1 K)^2 a_0 + (1 + a_1 K) a_1 - (1 + a_1 K) a_1 - (2 + a_1 K) a_0 a_1 K}{1 + 2a_0 K + a_1 K + a_0 a_1 K^2} \\
=&\frac{\left[(1 + a_1 K)^2 - (2 + a_1 K) a_1 K\right] a_0}{1 + 2a_0 K + a_1 K + a_0 a_1 K^2} \\
=&\frac{a_0}{1 + 2a_0 K + a_1 K + a_0 a_1 K^2} > 0.
\end{aligned}
\tag{A.166}
$$

Then, $Y_3 > 0$ for all $\beta \in (0, 1]$.

Now, we are going to show that $Y_2 > 0$ for all $\beta \in (0, 1]$:

$$
\begin{aligned}
Y_2 =& \left(\tfrac{1}{K} + a_1\right)\left(v_0^* + a_0\right) \tfrac{1}{K} + \left(\tfrac{1}{K} + a_1\right)\left(v_0^* + a_0\right)\left(\tfrac{1}{K} + v_0^*\right) a_0 K \\
&- \left(\tfrac{1}{K} + a_0\right) a_1 v_0^* - 2\left(\tfrac{1}{K} + a_0\right)(1 + a_1 K) a_0 v_0^* \\
=& \left(\tfrac{1}{K} + a_1\right) v_0^* \tfrac{1}{K} + \left(\tfrac{1}{K} + a_1\right) a_0 \tfrac{1}{K} + \left(\tfrac{1}{K} + a_1\right) a_0 K \left[v_0^{*2} + \left(\tfrac{1}{K} + a_0\right) v_0^* + a_0 \tfrac{1}{K}\right] \\
&- \left(\tfrac{1}{K} + a_0\right) a_1 v_0^* - 2\left(\tfrac{1}{K} + a_0\right)(1 + a_1 K) a_0 v_0^* \\
=& \left(\tfrac{1}{K} + a_1\right) v_0^* \tfrac{1}{K} + \left(\tfrac{1}{K} + a_1\right) a_0 \tfrac{1}{K} \\
&+ \left(\tfrac{1}{K} + a_1\right) a_0 K v_0^{*2} + \left(\tfrac{1}{K} + a_0\right)\left(\tfrac{1}{K} + a_1\right) a_0 K v_0^* + \left(\tfrac{1}{K} + a_1\right) a_0^2 \\
&- \left(\tfrac{1}{K} + a_0\right) a_1 v_0^* - 2\left(\tfrac{1}{K} + a_0\right)(1 + a_1 K) a_0 v_0^* \\
=& \left(\tfrac{1}{K} + a_1\right) a_0 K v_0^{*2} + \left(\tfrac{1}{K} + a_1\right) v_0^* \tfrac{1}{K} - \left(\tfrac{1}{K} + a_0\right) a_1 v_0^* \\
&+ \left(\tfrac{1}{K} + a_1\right) a_0 \tfrac{1}{K} + \left(\tfrac{1}{K} + a_1\right) a_0^2 \\
&+ \left(\tfrac{1}{K} + a_0\right)\left(\tfrac{1}{K} + a_1\right) a_0 K v_0^* - 2\left(\tfrac{1}{K} + a_0\right)(1 + a_1 K) a_0 v_0^* \\
=& \left(\tfrac{1}{K} + a_1\right) a_0 K v_0^{*2} + \left[\left(\tfrac{1}{K} + a_1\right) \tfrac{1}{K} - \left(\tfrac{1}{K} + a_0\right) a_1\right] v_0^* \\
&+ \left(\tfrac{1}{K} + a_0\right)\left(\tfrac{1}{K} + a_1\right) a_0 - \left(\tfrac{1}{K} + a_0\right)\left(\tfrac{1}{K} + a_1\right) a_0 K v_0^* \\
=& \left(\tfrac{1}{K} + a_1\right) a_0 K v_0^{*2} + \left(\tfrac{1}{K^2} - a_0 a_1\right) v_0^* + \left(\tfrac{1}{K} + a_0\right)\left(\tfrac{1}{K} + a_1\right) a_0 K \left(\tfrac{1}{K} - v_0^*\right) \\
=& \left[\left(\tfrac{1}{K} + a_1\right) a_0 K v_0^* + \left(\tfrac{1}{K^2} - a_0 a_1\right)\right] v_0^* + \left(\tfrac{1}{K} + a_0\right)\left(\tfrac{1}{K} + a_1\right)\left(\tfrac{1}{K} - v_0^*\right) a_0 K \\
=& Z_1 v_0^* + Z_2,
\end{aligned}
\tag{A.167}
$$

where

$$
Z_1 = \left(\tfrac{1}{K} + a_1\right) a_0 K v_0^* + \left(\tfrac{1}{K^2} - a_0 a_1\right) \tag{A.168}
$$

and

$$Z_2 = \left(\tfrac{1}{K} + a_0\right)\left(\tfrac{1}{K} + a_1\right)\left(\tfrac{1}{K} - \nu_0^*\right) a_0 K. \tag{A.169}$$

Given the values of a_0, a_1, ν_0^*, K, and (A.150), one has that $Z_2 > 0$ for all $\beta \in (0, 1]$, and $Z_1 = Z_1(\beta)$ is strictly decreasing with respect to β, because $\nu_0^*(\beta)$ strictly decreases by β, and $(a_1 + \frac{1}{K})a_0 K > 0$. Thus,

$$\begin{aligned}
Z_1 =& Z_1(\beta) \geq Z_1(1) = \left(\tfrac{1}{K} + a_1\right) a_0 K \nu_0^*(1) + \left(\tfrac{1}{K^2} - a_0 a_1\right) \\
=& \left(\tfrac{1}{K} + a_1\right) a_0 K \frac{a_0 + a_1 + a_0 a_1 K}{1 + 2a_0 K + a_1 K + a_0 a_1 K^2} + \left(\tfrac{1}{K^2} - a_0 a_1\right) \\
=& \frac{(a_0 + a_1 + a_0 a_1 K)\left(\tfrac{1}{K} + a_1\right) a_0 K + \left(1 + 2a_0 K + a_1 K + a_0 a_1 K^2\right)\left(\tfrac{1}{K^2} - a_0 a_1\right)}{1 + 2a_0 K + a_1 K + a_0 a_1 K^2} \\
=& \frac{(a_0 + a_1 + a_0 a_1 K)(1 + a_1 K) a_0 + \left(1 + 2a_0 K + a_1 K + a_0 a_1 K^2\right)\left(\tfrac{1}{K^2} - a_0 a_1\right)}{1 + 2a_0 K + a_1 K + a_0 a_1 K^2} \\
=& \frac{\left(a_0 + a_1 + 2a_0 a_1 K + a_1^2 K + a_0 a_1^2 K^2\right) a_0}{1 + 2a_0 K + a_1 K + a_0 a_1 K^2} \\
&+ \frac{\left(1 + 2a_0 K + a_1 K + a_0 a_1 K^2\right)\left(\tfrac{1}{K^2} - a_0 a_1\right)}{1 + 2a_0 K + a_1 K + a_0 a_1 K^2} \\
=& \frac{a_0^2 + \left(1 + 2a_0 K + a_1 K + a_0 a_1 K^2\right) a_0 a_1}{1 + 2a_0 K + a_1 K + a_0 a_1 K^2} \\
&+ \frac{\left(1 + 2a_0 K + a_1 K + a_0 a_1 K^2\right)\tfrac{1}{K^2} - \left(1 + 2a_0 K + a_1 K + a_0 a_1 K^2\right) a_0 a_1}{1 + 2a_0 K + a_1 K + a_0 a_1 K^2} \\
=& \frac{a_0^2 + \left(1 + 2a_0 K + a_1 K + a_0 a_1 K^2\right)\tfrac{1}{K^2}}{1 + 2a_0 K + a_1 K + a_0 a_1 K^2} \\
=& \frac{a_0^2}{1 + 2a_0 K + a_1 K + a_0 a_1 K^2} + \frac{1}{K^2} > 0.
\end{aligned} \tag{A.170}$$

Then, $Z_1 > 0$ for all $\beta \in (0, 1]$, which proves that $Y_2 = \nu_0^* Z_1 + Z_2 > 0$ for all $\beta \in (0, 1]$.

Now, since $Y_1, Y_2, Y_3 > 0$, we have that

$$T_1 = (1 - \beta)^2 Y_1 + (1 - \beta) Y_2 + \left(\tfrac{1}{K} + a_0\right) a_0 Y_3 > 0, \tag{A.171}$$

which proves that

$$X_1 = \frac{T_1}{T_2} > 0. \tag{A.172}$$

Since $X_1 > 0$ and $X_2 \geq 0$, then, if $X_3 \geq 0$ for $\beta_0 \in (0, 1]$, we have that

$$S_1 = X_1(G(b_1) + D) + X_2(G(b_0) + D) + X_3(b_1 - b_0) > 0, \tag{A.173}$$

for $\beta_0 \in (0, 1]$.

On the other hand, if $X_3 < 0$, for $\beta_0 \in (0, 1]$, then,

$$\begin{aligned}
S_1 =& X_1(G(b_1) + D) + X_2(G(b_0) + D) + X_3(b_1 - b_0) \\
=& \left[(1-\beta)\tfrac{1}{K} + a_0\right](v_0^* + a_0)\left(\tfrac{1}{K} + a_1\right)(G(b_1) + D) \\
& - \left[(1-\beta)v_0^* + a_0\right](v_1^* + a_1)\left(\tfrac{1}{K} + a_0\right)(G(b_1) + D) \\
& + X_2(G(b_0) + D) + X_3(b_1 - b_0) \\
=& (1-\beta)\left[\tfrac{1}{K}(v_0^* + a_0)\left(\tfrac{1}{K} + a_1\right) - v_0^*(v_1^* + a_1)\left(\tfrac{1}{K} + a_0\right)\right](G(b_1) + D) \\
& - a_0\left[(v_1^* + a_1)\left(\tfrac{1}{K} + a_0\right) - (v_0^* + a_0)\left(\tfrac{1}{K} + a_1\right)\right](G(b_1) + D) \\
& + X_2(G(b_0) + D) + X_3(b_1 - b_0) \\
=& (1-\beta)X_4(G(b_1) + D) - a_0X_3(G(b_1) + D) + X_2(G(b_0) + D) + X_3(b_1 - b_0) \\
=& (1-\beta)X_4(G(b_1) + D) - X_3\left[a_0(G(b_1) + D) - (b_1 - b_0)\right] + X_2(G(b_0) + D),
\end{aligned} \tag{A.174}$$

where

$$X_4 = \tfrac{1}{K}(v_0^* + a_0)\left(\tfrac{1}{K} + a_1\right) - v_0^*(v_1^* + a_1)\left(\tfrac{1}{K} + a_0\right). \tag{A.175}$$

Applying inequalities (A.150)–(A.175), we see that

$$\begin{aligned}
X_4 =& \tfrac{1}{K}(v_0^* + a_0)\left(\tfrac{1}{K} + a_1\right) - v_0^*(v_1^* + a_1)\left(\tfrac{1}{K} + a_0\right) \\
>& \tfrac{1}{K}(v_0^* + a_0)(v_1^* + a_1) - v_0^*(v_1^* + a_1)\left(\tfrac{1}{K} + a_0\right) \\
=& (v_1^* + a_1)\left[\tfrac{1}{K}(v_0^* + a_0) - v_0^*\left(\tfrac{1}{K} + a_0\right)\right] \\
=& (v_1^* + a_1)\left(a_0\tfrac{1}{K} - a_0v_0^*\right) \\
=& a_0(v_1^* + a_1)\left(\tfrac{1}{K} - v_0^*\right) > 0.
\end{aligned} \tag{A.176}$$

Thus, $X_4 > 0$ for $\beta_0 \in (0, 1]$, and since $X_2 \geq 0$, $X_3 < 0$ and assumption **A**3, we have that

$$\begin{aligned}
S_1 =& (1-\beta)X_4(G(b_1) + D) - X_3\left[a_0(G(b_1) + D) - (b_1 - b_0)\right] + X_2(G(b_0) + D) \\
\geq& - X_3\left[a_0(G(b_1) + D) - (b_1 - b_0)\right] > 0,
\end{aligned} \tag{A.177}$$

for $\beta_0 \in (0, 1]$.

Therefore, $S_1 > 0$ for all $\beta \in (0, 1]$, that is,

$$(p^c - p^*)(\beta) = \frac{S_1}{S_2} > 0, \tag{A.178}$$

which finally proves (A.140). The proof of the theorem is complete. ■

Theorem 2.7 *The functions $\pi_1^*(\beta)$ and $\pi_1^c(\beta)$ are strictly decreasing with respect to $\beta \in (0, 1]$. Moreover, the following inequalities hold:*

$$\pi_1^*(1) > \pi_1^c(1) \tag{A.179}$$

and

$$\lim_{\beta \to 0} \pi_1^*(\beta) < \lim_{\beta \to 0} \pi_1^c(\beta). \tag{A.180}$$

Proof First, we are going to show that π_1^* and π_1^c strictly decrease by β.

The function π_1^* is differentiable with respect to β and

$$\begin{aligned}
\pi_1^{*\prime} &= \left(p^* q_1^* - \frac{1}{2} a_1 q_1^{*2} - b_1 q_1^*\right)' = p^{*\prime} q_1^* + p^* q_1^{*\prime} - a_1 q_1^* q_1^{*\prime} - b_1 q_1^{*\prime} \\
&= p^{*\prime} q_1^* + \left(p^* - a_1 q_1^* - b_1\right) q_1^{*\prime} \\
&= p^{*\prime} q_1^* + \left(p^* - b_1 - a_1 \frac{p^* - b_1}{v_1^* + a_1}\right) q_1^{*\prime} \\
&= p^{*\prime} q_1^* + \left(1 - \frac{a_1}{v_1^* + a_1}\right) \left(p^* - b_1\right) q_1^{*\prime} \\
&= p^{*\prime} q_1^* + \frac{v_1^*}{v_1^* + a_1} \left(p^* - b_1\right) q_1^{*\prime}.
\end{aligned} \tag{A.181}$$

Given the values of a_1, b_1, v_1^*, p^*, q_1^*, $p^{*\prime}$ and $q_1^{*\prime}$, it's easy to see that

$$\pi_1^{*\prime} = p^{*\prime} q_1^* + \frac{v_1^*}{v_1^* + a_1} \left(p^* - b_1\right) q_1^{*\prime} < 0. \tag{A.182}$$

Similarly,

$$\pi_1^{c\prime} = p^{c\prime} q_1^c + \frac{\frac{1}{K}}{\frac{1}{K} + a_1} \left(p^c - b_1\right) q_1^{c\prime} < 0. \tag{A.183}$$

Because of that, π_1^* and π_1^c strictly decrease with respect to $\beta \in (0, 1]$.

Now consider the difference of the functions π_1^* and π_1^c as follows:

$$\begin{aligned}
\pi_1^c - \pi_1^* &= \left(p^c q_1^c - \frac{1}{2} a_1 q_1^{c2} - b_1 q_1^c\right) - \left(p^* q_1^* - \frac{1}{2} a_1 q_1^{*2} - b_1 q_1^*\right) \\
&= \left(p^c - b_1 - \frac{1}{2} a_1 q_1^c\right) q_1^c - \left(p^* - b_1 - \frac{1}{2} a_1 q_1^*\right) q_1^* \\
&= \left[\left(\frac{1}{K} + a_1\right) \frac{p^c - b_1}{\frac{1}{K} + a_1} - \frac{1}{2} a_1 q_1^c\right] q_1^c - \left[\left(v_1^* + a_1\right) \frac{p^* - b_1}{v_1^* + a_1} - \frac{1}{2} a_1 q_1^*\right] q_1^* \\
&= \left[\left(\frac{1}{K} + a_1\right) q_1^c - \frac{1}{2} a_1 q_1^c\right] q_1^c - \left[\left(v_1^* + a_1\right) q_1^* - \frac{1}{2} a_1 q_1^*\right] q_1^* \\
&= \left(\frac{1}{K} + \frac{1}{2} a_1\right) q_1^{c2} - \left(v_1^* + \frac{1}{2} a_1\right) q_1^{*2}.
\end{aligned} \tag{A.184}$$

From (A.97) we have that

$$
\begin{aligned}
q_1^* =& \frac{-(b_1-b_0)+\left[(1-\beta)v_0^*+a_0\right](G(b_1)+D)}{\left(v_0^*+a_0\right)+\left(v_1^*+a_1\right)\left\{1+\left[(1-\beta)v_0^*+a_0\right]K\right\}} \\
=& \frac{\left[(1-\beta)v_0^*+a_0\right](G(b_1)+D)-(b_1-b_0)}{\left(v_0^*+a_0\right)+\left(v_1^*+a_1\right)+\left[(1-\beta)v_0^*+a_0\right]\left(v_1^*+a_1\right)K},
\end{aligned}
\tag{A.185}
$$

and similarly to (A.97) and (A.185),

$$
\begin{aligned}
q_1^c =& \frac{p^c-b_1}{\frac{1}{K}+a_1} = \frac{\frac{\left(\frac{1}{K}+a_0\right)b_1+\left(\frac{1}{K}+a_1\right)b_0+\left[(1-\beta)\frac{1}{K}+a_0\right]\left(\frac{1}{K}+a_1\right)(T+D)}{\left(\frac{1}{K}+a_0\right)+\left(\frac{1}{K}+a_1\right)+\left[(1-\beta)\frac{1}{K}+a_0\right]\left(\frac{1}{K}+a_1\right)K}-b_1}{\frac{1}{K}+a_1} \\
=& \frac{\left(\frac{1}{K}+a_0\right)b_1+\left(\frac{1}{K}+a_1\right)b_0+\left[(1-\beta)\frac{1}{K}+a_0\right]\left(\frac{1}{K}+a_1\right)(T+D)}{\left(\frac{1}{K}+a_1\right)\left[\left(\frac{1}{K}+a_0\right)+\left(\frac{1}{K}+a_1\right)+\left[(1-\beta)\frac{1}{K}+a_0\right]\left(\frac{1}{K}+a_1\right)K\right]} \\
&- \frac{\left[\left(\frac{1}{K}+a_0\right)+\left(\frac{1}{K}+a_1\right)+\left[(1-\beta)\frac{1}{K}+a_0\right]\left(\frac{1}{K}+a_1\right)K\right]b_1}{\left(\frac{1}{K}+a_1\right)\left[\left(\frac{1}{K}+a_0\right)+\left(\frac{1}{K}+a_1\right)+\left[(1-\beta)\frac{1}{K}+a_0\right]\left(\frac{1}{K}+a_1\right)K\right]} \\
=& \frac{-\left(\frac{1}{K}+a_1\right)(b_1-b_0)+\left[(1-\beta)\frac{1}{K}+a_0\right]\left(\frac{1}{K}+a_1\right)(-Kb_1+T+D)}{\left(\frac{1}{K}+a_1\right)\left[\left(\frac{1}{K}+a_0\right)+\left(\frac{1}{K}+a_1\right)+\left[(1-\beta)\frac{1}{K}+a_0\right]\left(\frac{1}{K}+a_1\right)K\right]} \\
=& \frac{-(b_1-b_0)+\left[(1-\beta)\frac{1}{K}+a_0\right](-Kb_1+T+D)}{\left(\frac{1}{K}+a_0\right)+\left(\frac{1}{K}+a_1\right)+\left[(1-\beta)\frac{1}{K}+a_0\right]\left(\frac{1}{K}+a_1\right)K} \\
=& \frac{\left[(1-\beta)\frac{1}{K}+a_0\right](G(b_1)+D)-(b_1-b_0)}{\left(\frac{1}{K}+a_0\right)+\left(\frac{1}{K}+a_1\right)+\left[(1-\beta)\frac{1}{K}+a_0\right]\left(\frac{1}{K}+a_1\right)K}.
\end{aligned}
\tag{A.186}
$$

By substituting the expression of v_1^* given by (A.42) in Eq. (A.185) we have that

$$
\begin{aligned}
q_1^* =& \frac{\left[(1-\beta)v_0^*+a_0\right](G(b_1)+D)-(b_1-b_0)}{\left(v_0^*+a_0\right)+\left(v_1^*+a_1\right)+\left[(1-\beta)v_0^*+a_0\right]\left(v_1^*+a_1\right)K} \\
=& \frac{\left[(1-\beta)v_0^*+a_0\right](G(b_1)+D)-(b_1-b_0)}{\left(v_0^*+a_0\right)+\left(\frac{(1-\beta)v_0^*+a_0}{1+\left[(1-\beta)v_0^*+a_0\right]K}+a_1\right)+\left[(1-\beta)v_0^*+a_0\right]\left(\frac{(1-\beta)v_0^*+a_0}{1+\left[(1-\beta)v_0^*+a_0\right]K}+a_1\right)K} \\
=& \frac{\left(1+\left[(1-\beta)v_0^*+a_0\right]K\right)\left\{\left[(1-\beta)v_0^*+a_0\right](G(b_1)+D)-(b_1-b_0)\right\}}{\left(1+\left[(1-\beta)v_0^*+a_0\right]K\right)\left\{\left(v_0^*+a_0\right)+a_1+(1+a_1K)\left[(1-\beta)v_0^*+a_0\right]\right\}} \\
=& \frac{\left[(1-\beta)v_0^*+a_0\right](G(b_1)+D)-(b_1-b_0)}{\left(v_0^*+a_0+a_1\right)+(1+a_1K)\left[(1-\beta)v_0^*+a_0\right]}.
\end{aligned}
\tag{A.187}
$$

By Eq. (A.42),

$$
v_1^* = \frac{(1-\beta)v_0^*+a_0}{1+\left[(1-\beta)v_0^*+a_0\right]K},
$$

therefore,

$$
\begin{aligned}
\nu_1^* + \frac{1}{2}a_1 &= \frac{(1-\beta)\nu_0^* + a_0}{1 + \left[(1-\beta)\nu_0^* + a_0\right] K} + \frac{1}{2}a_1 \\
&= \frac{(1-\beta)\nu_0^* + a_0 + \frac{1}{2}a_1\left(1 + \left[(1-\beta)\nu_0^* + a_0\right] K\right)}{1 + \left[(1-\beta)\nu_0^* + a_0\right] K} \\
&= \frac{\frac{1}{2}a_1 + \left(1 + \frac{1}{2}a_1 K\right)\left[(1-\beta)\nu_0^* + a_0\right]}{1 + \left[(1-\beta)\nu_0^* + a_0\right] K}.
\end{aligned}
\tag{A.188}
$$

On the other hand, from the expression for q_1^c obtained from (A.186) we have that

$$
\begin{aligned}
q_1^c &= \frac{\left[(1-\beta)\frac{1}{K} + a_0\right](G(b_1) + D) - (b_1 - b_0)}{\left(\frac{1}{K} + a_0\right) + \left(\frac{1}{K} + a_1\right) + \left[(1-\beta)\frac{1}{K} + a_0\right]\left(\frac{1}{K} + a_1\right) K} \\
&= \frac{\left[(1-\beta)\frac{1}{K} + a_0\right](G(b_1) + D) - (b_1 - b_0)}{\frac{1}{K}(1 + a_0 K) + \frac{1}{K}(1 + a_1 K) + \frac{1}{K}[(1-\beta) + a_0 K](1 + a_1 K)} \\
&= \frac{\left\{\left[(1-\beta)\frac{1}{K} + a_0\right](G(b_1) + D) - (b_1 - b_0)\right\} K}{(1 + a_0 K) + (1 + a_1 K)[(2-\beta) + a_0 K]}.
\end{aligned}
\tag{A.189}
$$

Plugging Eqs. (A.187), (A.188) and (A.189) in Eq. (A.184) we deduce

$$
\begin{aligned}
\pi_1^c - \pi_1^* &= \left(\tfrac{1}{K} + \tfrac{1}{2}a_1\right) {q_1^c}^2 - \left(\nu_1^* + \tfrac{1}{2}a_1\right) {q_1^*}^2 \\
&= \left(\tfrac{1}{K} + \tfrac{1}{2}a_1\right)\left(\tfrac{\left\{\left[(1-\beta)\frac{1}{K}+a_0\right](G(b_1)+D)-(b_1-b_0)\right\}K}{(1+a_0K)+(1+a_1K)[(2-\beta)+a_0K]}\right)^2 \\
&\quad - \left(\tfrac{\frac{1}{2}a_1+\left(1+\frac{1}{2}a_1K\right)\left[(1-\beta)\nu_0^*+a_0\right]}{1+\left[(1-\beta)\nu_0^*+a_0\right]K}\right)\left(\tfrac{\left[(1-\beta)\nu_0^*+a_0\right](G(b_1)+D)-(b_1-b_0)}{\left(\nu_0^*+a_0+a_1\right)+(1+a_1K)\left[(1-\beta)\nu_0^*+a_0\right]}\right)^2 \\
&= \tfrac{1}{2}K\,(2 + a_1 K)\left(\tfrac{\left[(1-\beta)\frac{1}{K}+a_0\right](G(b_1)+D)-(b_1-b_0)}{(1+a_0K)+(1+a_1K)[(2-\beta)+a_0K]}\right)^2 \\
&\quad - \tfrac{1}{2}\left(\tfrac{a_1+(2+a_1K)\left[(1-\beta)\nu_0^*+a_0\right]}{1+\left[(1-\beta)\nu_0^*+a_0\right]K}\right)\left(\tfrac{\left[(1-\beta)\nu_0^*+a_0\right](G(b_1)+D)-(b_1-b_0)}{\left(\nu_0^*+a_0+a_1\right)+(1+a_1K)\left[(1-\beta)\nu_0^*+a_0\right]}\right)^2.
\end{aligned}
\tag{A.190}
$$

Then, to prove the inequalities (A.5) and (A.6) the following conditions has to be met:

$$
\pi_1^c(1) - \pi_1^*(1) = (\pi_1^c - \pi_1^*)(1) < 0 \tag{A.191}
$$

and

$$
\lim_{\beta\to 0} \pi_1^c(\beta) - \lim_{\beta\to 0} \pi_1^*(\beta) = \lim_{\beta\to 0} (\pi_1^c - \pi_1^*)(\beta) > 0. \tag{A.192}
$$

Evaluating the expression of ν_0^*, given by (A.41), for $\beta = 1$ and using the notation $\overline{\nu_0^*} = \nu_0^*(1)$, one has

$$\overline{v_0^*} = v_0^*(1) = \frac{2\left(a_0 + a_1 + a_0 a_1 K\right)}{\left(1 + 2a_0 K + a_1 K + a_0 a_1 K^2\right) + \sqrt{\left(1 + 2a_0 K + a_1 K + a_0 a_1 K^2\right)^2}}$$
$$= \frac{a_0 + a_1 + a_0 a_1 K}{1 + 2a_0 K + a_1 K + a_0 a_1 K^2}. \tag{A.193}$$

Now, we evaluate (A.190) for $\beta = 1$ to obtain

$$\begin{aligned}
(\pi_1^c - \pi_1^*)(1) =& \frac{1}{2} K \left(2 + a_1 K\right) \left(\frac{[a_0]\left(G(b_1) + D\right) - (b_1 - b_0)}{(1 + a_0 K) + (1 + a_1 K)\left[1 + a_0 K\right]} \right)^2 \\
& - \frac{1}{2} \left(\frac{a_1 + (2 + a_1 K)[a_0]}{1 + [a_0] K} \right) \left(\frac{[a_0]\left(G(b_1) + D\right) - (b_1 - b_0)}{\left(\overline{v_0^*} + a_0 + a_1\right) + (1 + a_1 K)[a_0]} \right)^2 \\
=& \frac{1}{2} K \left(2 + a_1 K\right) \left(\frac{a_0\left(G(b_1) + D\right) - (b_1 - b_0)}{(1 + a_0 K)(2 + a_1 K)} \right)^2 \\
& - \frac{1}{2} \left(\frac{a_1 + a_0 (2 + a_1 K)}{1 + a_0 K} \right) \left(\frac{a_0\left(G(b_1) + D\right) - (b_1 - b_0)}{\left(\overline{v_0^*} + a_0 + a_1\right) + a_0 (1 + a_1 K)} \right)^2 \\
=& \frac{1}{2} \frac{\left[a_0\left(G(b_1) + D\right) - (b_1 - b_0)\right]^2}{1 + a_0 K} \frac{K}{(1 + a_0 K)(2 + a_1 K)} \\
& - \frac{1}{2} \frac{\left[a_0\left(G(b_1) + D\right) - (b_1 - b_0)\right]^2}{1 + a_0 K} \frac{a_1 + a_0 (2 + a_1 K)}{\left[\left(\overline{v_0^*} + a_0 + a_1\right) + a_0 (1 + a_1 K)\right]^2} \\
=& \frac{1}{2} \frac{\left[a_0\left(G(b_1) + D\right) - (b_1 - b_0)\right]^2}{1 + a_0 K} \frac{K}{2 + 2a_0 K + a_1 K + a_0 a_1 K^2} \\
& - \frac{1}{2} \frac{\left[a_0\left(G(b_1) + D\right) - (b_1 - b_0)\right]^2}{1 + a_0 K} \frac{2a_0 + a_1 + a_0 a_1 K}{\left(\overline{v_0^*} + 2a_0 + a_1 + a_0 a_1 K\right)^2} \\
=& U_1 \frac{V_1}{W_1},
\end{aligned} \tag{A.194}$$

where

$$U_1 = \frac{1}{2} \frac{\left[a_0\left(G(b_1) + D\right) - (b_1 - b_0)\right]^2}{1 + a_0 K}, \tag{A.195}$$

$$V_1 = K \left(\overline{v_0^*} + 2a_0 + a_1 + a_0 a_1 K\right)^2 - \left(2a_0 + a_1 + a_0 a_1 K\right)\left(2 + 2a_0 K + a_1 K + a_0 a_1 K^2\right) \tag{A.196}$$

and

$$W_1 = \left(2 + 2a_0 K + a_1 K + a_0 a_1 K^2\right)\left(\overline{v_0^*} + 2a_0 + a_1 + a_0 a_1 K\right)^2. \tag{A.197}$$

Given the values of a_0, a_1, $\overline{v_0^*}$ and K, it isn't difficult to see that $U_1 > 0$ and $W_1 > 0$. Hence, to prove (A.191) it is enough to show that $V_1 > 0$. Indeed, plugging the expression of $\overline{v_0^*}$ given by (A.193) in (A.196), we have that

$$
\begin{aligned}
V_1 =& K\left(\widehat{\nu_0^*} + 2a_0 + a_1 + a_0a_1K\right)^2 - (2a_0 + a_1 + a_0a_1K)\left(2 + 2a_0K + a_1K + a_0a_1K^2\right) \\
=& K\left(\frac{a_0 + a_1 + a_0a_1K}{1 + 2a_0K + a_1K + a_0a_1K^2} + 2a_0 + a_1 + a_0a_1K\right)^2 \\
& - (2a_0 + a_1 + a_0a_1K)\left(2 + 2a_0K + a_1K + a_0a_1K^2\right) \\
=& K\left[a_0 + a_1 + a_0a_1K + (2a_0 + a_1 + a_0a_1K)\left(1 + 2a_0K + a_1K + a_0a_1K^2\right)\right]^2 \\
& - (2a_0 + a_1 + a_0a_1K)\left(2 + 2a_0K + a_1K + a_0a_1K^2\right)\left(1 + 2a_0K + a_1K + a_0a_1K^2\right)^2 \\
<& K\left[(2a_0 + a_1 + a_0a_1K) + (2a_0 + a_1 + a_0a_1K)\left(1 + 2a_0K + a_1K + a_0a_1K^2\right)\right]^2 \\
& - (2a_0 + a_1 + a_0a_1K)\left(2 + 2a_0K + a_1K + a_0a_1K^2\right)\left(1 + 2a_0K + a_1K + a_0a_1K^2\right)^2 \\
=& K\left[(2a_0 + a_1 + a_0a_1K)\left(2 + 2a_0K + a_1K + a_0a_1K^2\right)\right]^2 \\
& - (2a_0 + a_1 + a_0a_1K)\left(2 + 2a_0K + a_1K + a_0a_1K^2\right)\left(1 + 2a_0K + a_1K + a_0a_1K^2\right)^2 \\
=& \left[K(2a_0 + a_1 + a_0a_1K)\left(2 + 2a_0K + a_1K + a_0a_1K^2\right)\right. \\
& \left. - \left(1 + 2a_0K + a_1K + a_0a_1K^2\right)^2\right](2a_0 + a_1 + a_0a_1K)\left(2 + 2a_0K + a_1K + a_0a_1K^2\right) \\
=& \left[\left(2a_0K + a_1K + a_0a_1K^2\right)\left(2 + 2a_0K + a_1K + a_0a_1K^2\right)\right. \\
& \left. - \left(1 + 2a_0K + a_1K + a_0a_1K^2\right)^2\right](2a_0 + a_1 + a_0a_1K)\left(2 + 2a_0K + a_1K + a_0a_1K^2\right) \\
=& \left[\left(1 + 2a_0K + a_1K + a_0a_1K^2 - 1\right)\left(1 + 2a_0K + a_1K + a_0a_1K^2 + 1\right)\right. \\
& \left. - \left(1 + 2a_0K + a_1K + a_0a_1K^2\right)^2\right](2a_0 + a_1 + a_0a_1K)\left(2 + 2a_0K + a_1K + a_0a_1K^2\right) \\
=& \left[\left(1 + 2a_0K + a_1K + a_0a_1K^2\right)^2 - 1\right. \\
& \left. - \left(1 + 2a_0K + a_1K + a_0a_1K^2\right)^2\right](2a_0 + a_1 + a_0a_1K)\left(2 + 2a_0K + a_1K + a_0a_1K^2\right) \\
=& -(2a_0 + a_1 + a_0a_1K)\left(2 + 2a_0K + a_1K + a_0a_1K^2\right) < 0.
\end{aligned}
\tag{A.198}
$$

Therefore $V_1 < 0$. Then, since $U_1 > 0$ and $W_1 > 0$, we have that

$$
(\pi_1^c - \pi_1^*)(1) = U_1\frac{V_1}{W_1} < 0, \tag{A.199}
$$

which proves (A.191).

Now, we need only to prove (A.192). Using the notation $\widehat{\nu_0^*} = \lim_{\beta \to 0} \nu_0^*(\beta)$ given by (A.141), from (A.190) we have that

$$
\begin{aligned}
&\lim_{\beta\to 0}(\pi_1^c - \pi_1^*)(\beta) = \\
&= \frac{1}{2}K\,(2+a_1K)\left(\frac{\left[\frac{1}{K}+a_0\right](G(b_1)+D)-(b_1-b_0)}{(1+a_0K)+(1+a_1K)\,[2+a_0K]}\right)^2 \\
&-\frac{1}{2}\left(\frac{a_1+(2+a_1K)\left[\widehat{\nu_0^*}+a_0\right]}{1+\left[\widehat{\nu_0^*}+a_0\right]K}\right)\left(\frac{\left[\widehat{\nu_0^*}+a_0\right](G(b_1)+D)-(b_1-b_0)}{\left(\widehat{\nu_0^*}+a_0+a_1\right)+(1+a_1K)\left[\widehat{\nu_0^*}+a_0\right]}\right)^2 \\
&= \frac{1}{2}K\,(2+a_1K)\left(\frac{\left(\frac{1}{K}+a_0\right)(G(b_1)+D)-(b_1-b_0)}{(1+a_0K)+(1+a_1K)+(1+a_0K)\,(1+a_1K)}\right)^2 \\
&-\frac{1}{2}\left(\frac{a_1+(2+a_1K)\left(\widehat{\nu_0^*}+a_0\right)}{1+\left(\widehat{\nu_0^*}+a_0\right)K}\right)\left(\frac{\left(\widehat{\nu_0^*}+a_0\right)(G(b_1)+D)-(b_1-b_0)}{a_1+(2+a_1K)\left(\widehat{\nu_0^*}+a_0\right)}\right)^2 \\
&= \frac{1}{2}K\,(2+a_1K)\left(\frac{\left(\frac{1}{K}+a_0\right)(G(b_1)+D)-(b_1-b_0)}{(1+a_1K)+(1+a_0K)\,(2+a_1K)}\right)^2 \\
&-\frac{1}{2}\frac{1}{1+\left(\widehat{\nu_0^*}+a_0\right)K}\frac{\left[\left(\widehat{\nu_0^*}+a_0\right)(G(b_1)+D)-(b_1-b_0)\right]^2}{a_1+(2+a_1K)\left(\widehat{\nu_0^*}+a_0\right)} \\
&= \frac{1}{2}\frac{V_2}{W_2},
\end{aligned}
\tag{A.200}
$$

where

$$
\begin{aligned}
V_2 =& K\,(2+a_1K)\left[\left(\frac{1}{K}+a_0\right)(G(b_1)+D)-(b_1-b_0)\right]^2\times \\
&\left[1+\left(\widehat{\nu_0^*}+a_0\right)K\right]\left[a_1+(2+a_1K)\left(\widehat{\nu_0^*}+a_0\right)\right] \\
&-[(1+a_1K)+(1+a_0K)\,(2+a_1K)]^2\times \\
&\left[\left(\widehat{\nu_0^*}+a_0\right)(G(b_1)+D)-(b_1-b_0)\right]^2
\end{aligned}
\tag{A.201}
$$

and

$$
\begin{aligned}
W_2 =& [(1+a_1K)+(1+a_0K)\,(2+a_1K)]^2\times \\
&\left[1+\left(\widehat{\nu_0^*}+a_0\right)K\right]\left[a_1+(2+a_1K)\left(\widehat{\nu_0^*}+a_0\right)\right].
\end{aligned}
\tag{A.202}
$$

For arbitrary fixed values of a_0, a_1, $\widehat{\nu_0^*}$ and K, it is evident that $W_2 > 0$. Hence, to prove (A.192) it lacks only to show that $V_2 > 0$. Indeed,

$$
\begin{aligned}
V_2 =& \left(\left[(1+a_1K)+(2+a_1K)\left(\widehat{\nu_0^*}+a_0\right)K\right]^2-1\right)\left[\left(\tfrac{1}{K}+a_0\right)(G(b_1)+D)-(b_1-b_0)\right]^2 \\
&-[(1+a_1K)+(1+a_0K)\,(2+a_1K)]^2\left[\left(\widehat{\nu_0^*}+a_0\right)(G(b_1)+D)-(b_1-b_0)\right]^2 \\
=& \left[(1+a_1K)+\left(\widehat{\nu_0^*}+a_0\right)(2+a_1K)\,K\right]^2\left[\left(\tfrac{1}{K}+a_0\right)(G(b_1)+D)-(b_1-b_0)\right]^2
\end{aligned}
$$

$$
\begin{aligned}
&-\left[(1+a_1K)+\left(\tfrac{1}{K}+a_0\right)(2+a_1K)\,K\right]^2\left[\left(\widehat{\nu_0^*}+a_0\right)(G(b_1)+D)-(b_1-b_0)\right]^2 \\
&-\left[\left(\tfrac{1}{K}+a_0\right)(G(b_1)+D)-(b_1-b_0)\right]^2.
\end{aligned} \tag{A.203}
$$

Now introduce the following notation:

$$\eta = 1 + a_1 K > 0, \tag{A.204}$$

$$\xi = K(1+\eta) = K(2+a_1K) > 0, \tag{A.205}$$

$$\mathscr{Z} = \eta + a_0\xi = (1+a_1K) + a_0K(2+a_1K) > 0, \tag{A.206}$$

$$G_1 = G(b_1) + D > 0 \tag{A.207}$$

and

$$G_3 = a_0G_1 - (b_1 - b_0) = a_0(G(b_1)+D) - (b_1-b_0) > 0. \tag{A.208}$$

Based on that, we can rewrite (A.203) as follows:

$$
\begin{aligned}
V_2 =& \left[(1+a_1K)+\left(\widehat{\nu_0^*}+a_0\right)(2+a_1K)\,K\right]^2\left[\left(\tfrac{1}{K}+a_0\right)(G(b_1)+D)-(b_1-b_0)\right]^2 \\
&-\left[(1+a_1K)+\left(\tfrac{1}{K}+a_0\right)(2+a_1K)\,K\right]^2\left[\left(\widehat{\nu_0^*}+a_0\right)(G(b_1)+D)-(b_1-b_0)\right]^2 \\
&-\left[\left(\tfrac{1}{K}+a_0\right)(G(b_1)+D)-(b_1-b_0)\right]^2 \\
=& \left(\widehat{\nu_0^*}\xi+\mathscr{Z}\right)^2\left(\tfrac{1}{K}G_1+G_3\right)^2-\left(\tfrac{1}{K}\xi+\mathscr{Z}\right)^2\left(\widehat{\nu_0^*}G_1+G_3\right)^2-\left(\tfrac{1}{K}G_1+G_3\right)^2 \\
=& \left(\tfrac{1}{K}\widehat{\nu_0^*}\xi G_1+\tfrac{1}{K}\mathscr{Z}G_1+\widehat{\nu_0^*}\xi G_3+\mathscr{Z}G_3\right)^2-\left(\tfrac{1}{K}\widehat{\nu_0^*}\xi G_1+\widehat{\nu_0^*}\mathscr{Z}G_1+\tfrac{1}{K}\xi G_3+\mathscr{Z}G_3\right)^2 \\
&-\left(\frac{1}{K^2}G_1^2+2\tfrac{1}{K}G_1G_3+G_3^2\right) \\
=& \left[2\left(\tfrac{1}{K}\widehat{\nu_0^*}\xi G_1+\mathscr{Z}G_3\right)+\left(\tfrac{1}{K}+\widehat{\nu_0^*}\right)(\mathscr{Z}G_1+\xi G_3)\right]\left[\left(\tfrac{1}{K}-\widehat{\nu_0^*}\right)(\mathscr{Z}G_1-\xi G_3)\right] \\
&-\left(\frac{1}{K^2}G_1^2+2\tfrac{1}{K}G_1G_3\right)-G_3^2 \\
=& 2\left(\tfrac{1}{K}\widehat{\nu_0^*}\xi G_1+\mathscr{Z}G_3\right)\left(\tfrac{1}{K}-\widehat{\nu_0^*}\right)(\mathscr{Z}G_1-\xi G_3)+\left(\tfrac{1}{K^2}-\widehat{\nu_0^*}^2\right)\left(\mathscr{Z}^2G_1^2-\xi^2G_3^2\right) \\
&-\tfrac{1}{K}G_1\left(\tfrac{1}{K}G_1+2G_3\right)-G_3^2 \\
=& \left(\tfrac{1}{K^2}-\widehat{\nu_0^*}^2\right)\left(\mathscr{Z}^2G_1^2-\xi^2G_3^2\right)-\tfrac{1}{K}G_1\left(\tfrac{1}{K}G_1+2G_3\right) \\
&+2\mathscr{Z}G_3\left(\tfrac{1}{K}-\widehat{\nu_0^*}\right)(\mathscr{Z}G_1-\xi G_3)-G_3^2 \\
&+2\tfrac{1}{K}\widehat{\nu_0^*}\xi G_1\left(\tfrac{1}{K}-\widehat{\nu_0^*}\right)(\mathscr{Z}G_1-\xi G_3) \\
=& \mathscr{P}_1+\mathscr{Q}_1+\mathscr{R}_1,
\end{aligned} \tag{A.209}
$$

where

$$\mathscr{P}_1 = \left(\tfrac{1}{K^2} - \widehat{\nu_0^*}^2\right)\left(\mathscr{L}^2 G_1^2 - \xi^2 G_3^2\right) - \tfrac{1}{K} G_1 \left(\tfrac{1}{K} G_1 + 2G_3\right), \tag{A.210}$$

$$\mathscr{Q}_1 = 2\mathscr{L} G_3 \left(\tfrac{1}{K} - \widehat{\nu_0^*}\right)\left(\mathscr{L} G_1 - \xi G_3\right) - G_3^2 \tag{A.211}$$

and

$$\mathscr{R}_1 = 2\tfrac{1}{K}\widehat{\nu_0^*}\xi G_1 \left(\tfrac{1}{K} - \widehat{\nu_0^*}\right)\left(\mathscr{L} G_1 - \xi G_3\right). \tag{A.212}$$

Now, we are going to show that

$$\mathscr{L} G_1 - \xi G_3 > 0. \tag{A.213}$$

Using (A.206) and (A.208) we have that

$$\begin{aligned}\mathscr{L} G_1 - \xi G_3 &= \left(\eta + a_0\xi\right) G_1 - \xi\left(a_0 G_1 - (b_1 - b_0)\right)\\ &= \eta G_1 + \xi(b_1 - b_0) \geq \eta G_1 > 0,\end{aligned} \tag{A.214}$$

which proves (A.213).

Thus, given the values of $\widehat{\nu_0^*}$, K, Eqs. (A.205), (A.207), (A.150) and (A.213), we can conclude that $\mathscr{R}_1 > 0$.

Now,

$$\begin{aligned}\mathscr{Q}_1 &= 2\mathscr{L} G_3 \left(\tfrac{1}{K} - \widehat{\nu_0^*}\right)\left(\mathscr{L} G_1 - \xi G_3\right) - G_3^2\\ &= \left[2\mathscr{L}\left(\tfrac{1}{K} - \widehat{\nu_0^*}\right)\left(\mathscr{L} G_1 - \xi G_3\right) - G_3\right] G_3,\end{aligned} \tag{A.215}$$

and using (A.208) we can rewrite (A.215)as follows:

$$\begin{aligned}\mathscr{Q}_1 &= \left[2\mathscr{L}\left(\tfrac{1}{K} - \widehat{\nu_0^*}\right)\left(\mathscr{L} G_1 - \xi G_3\right) - G_3\right] G_3\\ &= \left\{2\mathscr{L}\left(\tfrac{1}{K} - \widehat{\nu_0^*}\right)\left(\mathscr{L} G_1 - \xi\left[a_0 G_1 - (b_1 - b_0)\right]\right) - \left[a_0 G_1 - (b_1 - b_0)\right]\right\} G_3\\ &= \left\{2\mathscr{L}\left(\tfrac{1}{K} - \widehat{\nu_0^*}\right)\left[\mathscr{L} G_1 - a_0\xi G_1 + \xi(b_1 - b_0)\right] - a_0 G_1 + (b_1 - b_0)\right\} G_3\\ &= \left\{2\mathscr{L}\left(\tfrac{1}{K} - \widehat{\nu_0^*}\right)\left[\left(\mathscr{L} - a_0\xi\right) G_1 + \xi(b_1 - b_0)\right] - a_0 G_1 + (b_1 - b_0)\right\} G_3\\ &= \left\{2\mathscr{L}\left(\mathscr{L} - a_0\xi\right)\left(\tfrac{1}{K} - \widehat{\nu_0^*}\right) G_1 + 2\xi\mathscr{L}\left(\tfrac{1}{K} - \widehat{\nu_0^*}\right)(b_1 - b_0) - a_0 G_1 + (b_1 - b_0)\right\} G_3\\ &= \left\{\left[2\mathscr{L}\left(\mathscr{L} - a_0\xi\right)\left(\tfrac{1}{K} - \widehat{\nu_0^*}\right) - a_0\right] G_1 + \left[2\xi\mathscr{L}\left(\tfrac{1}{K} - \widehat{\nu_0^*}\right) + 1\right](b_1 - b_0)\right\} G_3.\end{aligned} \tag{A.216}$$

Moreover, from (A.206), we have that

$$\eta = \mathscr{L} - a_0\xi = 1 + a_1 K. \tag{A.217}$$

Substituting (A.217) in (A.216) we have that

$$
\begin{aligned}
\mathcal{Q}_1 &= \left\{\left[2\mathcal{Z}\left(\mathcal{Z}-a_0\xi\right)\left(\tfrac{1}{K}-\widehat{\nu_0^*}\right)-a_0\right]G_1+\left[2\xi\mathcal{Z}\left(\tfrac{1}{K}-\widehat{\nu_0^*}\right)+1\right](b_1-b_0)\right\}G_3\\
&= \left\{\left[2\mathcal{Z}\left(1+a_1K\right)\left(\tfrac{1}{K}-\widehat{\nu_0^*}\right)-a_0\right]G_1+\left[2\xi\mathcal{Z}\left(\tfrac{1}{K}-\widehat{\nu_0^*}\right)+1\right](b_1-b_0)\right\}G_3\\
&= \left\{\left[2a_1K\mathcal{Z}\left(\tfrac{1}{K}-\widehat{\nu_0^*}\right)+2\mathcal{Z}\left(\tfrac{1}{K}-\widehat{\nu_0^*}\right)-a_0\right]G_1+\left[2\xi\mathcal{Z}\left(\tfrac{1}{K}-\widehat{\nu_0^*}\right)+1\right](b_1-b_0)\right\}G_3\\
&= \left[\left(V_3+W_3\right)G_1+U_3(b_1-b_0)\right]G_3,
\end{aligned}
\tag{A.218}
$$

where

$$V_3 = 2a_1K\mathcal{Z}\left(\tfrac{1}{K}-\widehat{\nu_0^*}\right), \tag{A.219}$$

$$W_3 = 2\mathcal{Z}\left(\tfrac{1}{K}-\widehat{\nu_0^*}\right)-a_0 \tag{A.220}$$

and

$$U_3 = 2\xi\mathcal{Z}\left(\tfrac{1}{K}-\widehat{\nu_0^*}\right)+1. \tag{A.221}$$

For any given values of a_1, K, ξ and $\mathcal{Z}$, one easily deduces that $V_3 > 0$ and $U_3 > 0$. Now, we are going to show that $W_3 > 0$. In order to do that, we first substitute (A.206) in (A.220) to get:

$$
\begin{aligned}
W_3 =& 2\mathcal{Z}\left(\tfrac{1}{K}-\widehat{\nu_0^*}\right)-a_0\\
=& 2\left[(1+a_1K)+a_0K\left(2+a_1K\right)\right]\left(\tfrac{1}{K}-\widehat{\nu_0^*}\right)-a_0\\
>& a_0K\left(2+a_1K\right)\left(\tfrac{1}{K}-\widehat{\nu_0^*}\right)-a_0\\
=& a_0\left[K\left(2+a_1K\right)\left(\tfrac{1}{K}-\widehat{\nu_0^*}\right)-1\right].
\end{aligned}
\tag{A.222}
$$

Now, making use of the expression of $\widehat{\nu_0^*}$ given by (A.141) we have that

$$
\begin{aligned}
\tfrac{1}{K}-\widehat{\nu_0^*} =& \tfrac{1}{K}-\frac{2(a_0+a_1+a_0a_1K)}{(2a_0K+a_0a_1K^2)+\sqrt{\left(2a_0K+a_0a_1K^2\right)^2+4\left(2K+a_1K^2\right)(a_0+a_1+a_0a_1K)}}\\
=& \frac{(2a_0K+a_0a_1K^2)+\sqrt{\left(2a_0K+a_0a_1K^2\right)^2+4\left(2K+a_1K^2\right)(a_0+a_1+a_0a_1K)}-2K(a_0+a_1+a_0a_1K)}{K\left[(2a_0K+a_0a_1K^2)+\sqrt{\left(2a_0K+a_0a_1K^2\right)^2+4\left(2K+a_1K^2\right)(a_0+a_1+a_0a_1K)}\right]}\\
=& \frac{\sqrt{\left(2a_0K+a_0a_1K^2\right)^2+4\left(2K+a_1K^2\right)(a_0+a_1+a_0a_1K)}-\left(2a_1K+a_0a_1K^2\right)}{K\left[(2a_0K+a_0a_1K^2)+\sqrt{\left(2a_0K+a_0a_1K^2\right)^2+4\left(2K+a_1K^2\right)(a_0+a_1+a_0a_1K)}\right]}\\
=& \frac{\sqrt{\left(2a_0K+a_0a_1K^2\right)^2+4\left(2K+a_1K^2\right)(a_0+a_1+a_0a_1K)}-a_1K(2+a_0K)}{K\left[(2a_0K+a_0a_1K^2)+\sqrt{\left(2a_0K+a_0a_1K^2\right)^2+4\left(2K+a_1K^2\right)(a_0+a_1+a_0a_1K)}\right]}.
\end{aligned}
\tag{A.223}
$$

Now plugging (A.223) in (A.222) we get:

$$
\begin{aligned}
W_3 > {} & a_0\left[K\left(2+a_1K\right)\left(\tfrac{1}{K}-\widehat{\nu_0^*}\right)-1\right] \\
= {} & a_0\left[K\left(2+a_1K\right)\frac{\sqrt{\left(2a_0K+a_0a_1K^2\right)^2+4\left(2K+a_1K^2\right)\left(a_0+a_1+a_0a_1K\right)}-a_1K\left(2+a_0K\right)}{K\left[\left(2a_0K+a_0a_1K^2\right)+\sqrt{\left(2a_0K+a_0a_1K^2\right)^2+4\left(2K+a_1K^2\right)\left(a_0+a_1+a_0a_1K\right)}\right]}-1\right] \\
= {} & a_0\left[\left(2+a_1K\right)\frac{\sqrt{\left(2a_0K+a_0a_1K^2\right)^2+4\left(2K+a_1K^2\right)\left(a_0+a_1+a_0a_1K\right)}-a_1K\left(2+a_0K\right)}{\left(2a_0K+a_0a_1K^2\right)+\sqrt{\left(2a_0K+a_0a_1K^2\right)^2+4\left(2K+a_1K^2\right)\left(a_0+a_1+a_0a_1K\right)}}-1\right] \\
= {} & \frac{a_0}{\left(2a_0K+a_0a_1K^2\right)+\sqrt{\left(2a_0K+a_0a_1K^2\right)^2+4\left(2K+a_1K^2\right)\left(a_0+a_1+a_0a_1K\right)}}\Bigg[\\
& \left(2+a_1K\right)\sqrt{\left(2a_0K+a_0a_1K^2\right)^2+4\left(2K+a_1K^2\right)\left(a_0+a_1+a_0a_1K\right)} \\
& -a_1K\left(2+a_0K\right)\left(2+a_1K\right)-\left(2a_0K+a_0a_1K^2\right) \\
& -\sqrt{\left(2a_0K+a_0a_1K^2\right)^2+4\left(2K+a_1K^2\right)\left(a_0+a_1+a_0a_1K\right)}\Bigg] \\
= {} & \frac{a_0}{\left(2a_0K+a_0a_1K^2\right)+\sqrt{\left(2a_0K+a_0a_1K^2\right)^2+4\left(2K+a_1K^2\right)\left(a_0+a_1+a_0a_1K\right)}}\Bigg[\\
& \left[\left(2+a_1K\right)-1\right]\sqrt{\left(2a_0K+a_0a_1K^2\right)^2+4\left(2K+a_1K^2\right)\left(a_0+a_1+a_0a_1K\right)} \\
& -a_1K\left(2+a_0K\right)\left(2+a_1K\right)-a_0K\left(2+a_1K\right)\Bigg] \\
= {} & \frac{a_0}{\left(2a_0K+a_0a_1K^2\right)+\sqrt{\left(2a_0K+a_0a_1K^2\right)^2+4\left(2K+a_1K^2\right)\left(a_0+a_1+a_0a_1K\right)}}\Bigg\{ \\
& \left(1+a_1K\right)\sqrt{\left(2a_0K+a_0a_1K^2\right)^2+4\left(2K+a_1K^2\right)\left(a_0+a_1+a_0a_1K\right)} \\
& -K\left(2+a_1K\right)\left[a_1\left(2+a_0K\right)+a_0\right]\Bigg\} \\
= {} & \frac{a_0}{\left(2a_0K+a_0a_1K^2\right)+\sqrt{\left(2a_0K+a_0a_1K^2\right)^2+4\left(2K+a_1K^2\right)\left(a_0+a_1+a_0a_1K\right)}}\Bigg\{ \\
& \left(1+a_1K\right)\sqrt{\left(2a_0K+a_0a_1K^2\right)^2+4\left(2K+a_1K^2\right)\left(a_0+a_1+a_0a_1K\right)} \\
& -K\left(2+a_1K\right)\left[a_0\left(1+a_1K\right)+2a_1\right]\Bigg\} \\
= {} & a_0\frac{V_4}{W_4},
\end{aligned}
\tag{A.224}
$$

where

$$V_4 = (1 + a_1 K) \sqrt{\left(2a_0 K + a_0 a_1 K^2\right)^2 + 4\left(2K + a_1 K^2\right)(a_0 + a_1 + a_0 a_1 K)} \\ - K(2 + a_1 K)\left[a_0(1 + a_1 K) + 2a_1\right] \tag{A.225}$$

and

$$W_4 = \left(2a_0 K + a_0 a_1 K^2\right) + \sqrt{\left(2a_0 K + a_0 a_1 K^2\right)^2 + 4\left(2K + a_1 K^2\right)(a_0 + a_1 + a_0 a_1 K)}. \tag{A.226}$$

For any values of a_0, a_1 and K, we see that $W_4 > 0$, thus, to compute the value of (A.224) we need only to estimate the value of V_4. Suppose that

$$V_4 = (1 + a_1 K) \sqrt{\left(2a_0 K + a_0 a_1 K^2\right)^2 + 4\left(2K + a_1 K^2\right)(a_0 + a_1 + a_0 a_1 K)} \\ - K(2 + a_1 K)\left[a_0(1 + a_1 K) + 2a_1\right] \le 0. \tag{A.227}$$

Then, we would have

$$(1 + a_1 K) \sqrt{\left(2a_0 K + a_0 a_1 K^2\right)^2 + 4\left(2K + a_1 K^2\right)(a_0 + a_1 + a_0 a_1 K)} \\ \le K(2 + a_1 K)\left[a_0(1 + a_1 K) + 2a_1\right]. \tag{A.228}$$

Both sides of (A.228) are positive, then, by squaring them we have that

$$(1 + a_1 K)^2 \left[\left(2a_0 K + a_0 a_1 K^2\right)^2 + 4\left(2K + a_1 K^2\right)(a_0 + a_1 + a_0 a_1 K)\right] \\ \le K^2 (2 + a_1 K)^2 \left[a_0(1 + a_1 K) + 2a_1\right]^2. \tag{A.229}$$

Thus,

$$(1 + a_1 K)^2 \left[a_0^2 K^2 (2 + a_1 K)^2 + 4K(2 + a_1 K)(a_0 + a_1 + a_0 a_1 K)\right] \\ \le K^2 (2 + a_1 K)^2 \left[a_0(1 + a_1 K) + 2a_1\right]^2, \tag{A.230}$$

which leads to

$$K(2 + a_1 K)(1 + a_1 K)^2 \left[a_0^2 K(2 + a_1 K) + 4(a_0 + a_1 + a_0 a_1 K)\right] \\ \le K^2 (2 + a_1 K)^2 \left[a_0(1 + a_1 K) + 2a_1\right]^2, \tag{A.231}$$

where, $K(2 + a_1 K) > 0$. Therefore,

$$(1 + a_1 K)^2 \left[a_0^2 K(2 + a_1 K) + 4(a_0 + a_1 + a_0 a_1 K)\right] \\ \le K(2 + a_1 K)\left[a_0(1 + a_1 K) + 2a_1\right]^2. \tag{A.232}$$

Now, expanding the squares we find that:

$$\begin{aligned}&(1+a_1K)^2\left[a_0^2K\,(2+a_1K)+4\,(a_0+a_1+a_0a_1K)\right]\\&\le K\,(2+a_1K)\left[a_0^2\,(1+a_1K)^2+4a_0a_1\,(1+a_1K)+4a_1^2\right],\end{aligned}\tag{A.233}$$

and by distributing some terms:

$$\begin{aligned}&a_0^2K\,(2+a_1K)\,(1+a_1K)^2+4\,(a_0+a_1+a_0a_1K)\,(1+a_1K)^2\\&\le a_0^2K\,(2+a_1K)\,(1+a_1K)^2+4a_0a_1K\,(2+a_1K)\,(1+a_1K)+4a_1^2K\,(2+a_1K)\,.\end{aligned}\tag{A.234}$$

Thus,

$$\begin{aligned}&4\,(a_0+a_1+a_0a_1K)\,(1+a_1K)^2\\&\le 4a_0a_1K\,(2+a_1K)\,(1+a_1K)+4a_1^2K\,(2+a_1K)\,,\end{aligned}\tag{A.235}$$

and distributing again:

$$\begin{aligned}&4\,(a_0+a_1)\,(1+a_1K)^2+4a_0a_1K\,(1+a_1K)^2\\&\le 4a_0a_1K\,(1+a_1K)+4a_0a_1K\,(1+a_1K)^2+4a_1^2K\,(2+a_1K)\,,\end{aligned}\tag{A.236}$$

which leads to

$$\begin{aligned}&4\,(a_0+a_1)\,(1+a_1K)^2\\&\le 4a_0a_1K\,(1+a_1K)+4a_1^2K\,(2+a_1K)\,.\end{aligned}\tag{A.237}$$

Finally,

$$\begin{aligned}&(a_0+a_1)\,(1+a_1K)^2\\&\le a_0a_1K\,(1+a_1K)+a_1^2K\,(2+a_1K)\,,\end{aligned}\tag{A.238}$$

and by distributing some terms and expanding the squares again we deduce that

$$\begin{aligned}&(a_0+a_1)\left(1+2a_1K+a_1^2K^2\right)\\&\le a_0a_1K\,(1+a_1K)+a_1^2K\,(1+a_1K)+a_1^2K,\end{aligned}\tag{A.239}$$

thus,

$$\begin{aligned}&(a_0+a_1)\left(1+2a_1K+a_1^2K^2\right)\\&\le (a_0+a_1)\,a_1K\,(1+a_1K)+a_1^2K,\end{aligned}\tag{A.240}$$

which leads to

$$\begin{aligned}&(a_0+a_1)\left(1+2a_1K+a_1^2K^2\right)\\&\le (a_0+a_1)\left(a_1K+a_1^2K^2\right)+a_1^2K.\end{aligned}\tag{A.241}$$

Hence,

$$(a_0 + a_1)(1 + a_1 K) \leq a_1^2 K, \tag{A.242}$$

which finally leads to

$$a_0 + a_1 + a_0 a_1 K + a_1^2 K \leq a_1^2 K, \tag{A.243}$$

which is impossible since $a_0 + a_1 + a_0 a_1 K > 0$.

On a base of that, we conclude that

$$V_4 > 0, \tag{A.244}$$

so,

$$W_3 > a_0 \frac{V_4}{W_4} > 0. \tag{A.245}$$

Now, since $V_3 > 0$, $W_3 > 0$ and $U_3 > 0$, and given the values of b_0, b_1, G_1 and G_3, we have that

$$\mathscr{Q}_1 = [(V_3 + W_3) G_1 + U_3(b_1 - b_0)] G_3 > 0. \tag{A.246}$$

So we need only to estimate the value of $\mathscr{P}_1$.

$$\begin{aligned}
\mathscr{P}_1 &= \left(\tfrac{1}{K^2} - \widehat{\nu_0^*}^2\right)\left(\mathscr{Z}^2 G_1^2 - \xi^2 G_3^2\right) - \tfrac{1}{K} G_1 \left(\tfrac{1}{K} G_1 + 2G_3\right) \\
&= \left(\tfrac{1}{K} + \widehat{\nu_0^*}\right)(\mathscr{Z} G_1 + \xi G_3)\left(\tfrac{1}{K} - \widehat{\nu_0^*}\right)(\mathscr{Z} G_1 - \xi G_3) - \tfrac{1}{K} G_1 \left(\tfrac{1}{K} G_1 + 2G_3\right) \\
&= \left(\tfrac{1}{K}\mathscr{Z} G_1 + \tfrac{1}{K}\xi G_3 + \widehat{\nu_0^*}\mathscr{Z} G_1 + \widehat{\nu_0^*}\xi G_3\right)\left(\tfrac{1}{K} - \widehat{\nu_0^*}\right)(\mathscr{Z} G_1 - \xi G_3) \\
&\quad - \tfrac{1}{K} G_1 \left(\tfrac{1}{K} G_1 + 2G_3\right).
\end{aligned} \tag{A.247}$$

By Eqs. (A.205), (A.208) and (A.217), we have that

$$\xi = K(2 + a_1 K) > 2K, \tag{A.248}$$

$$\mathscr{Z} G_1 - \xi G_3 = \mathscr{Z} G_1 - \xi [a_0 G_1 - (b_1 - b_0)] \geq \mathscr{Z} G_1 - a_0 \xi G_1 = \eta G_1, \tag{A.249}$$

$$\mathscr{Z} = \eta + a_0 \xi > \eta \tag{A.250}$$

and

$$\widehat{\nu_0^*}\xi G_3 > 0. \tag{A.251}$$

Using inequalities (A.248)–(A.251) in (A.247) we get:

$$\begin{aligned}\mathscr{P}_1 &= \left(\tfrac{1}{K}\mathscr{Z}G_1 + \tfrac{1}{K}\xi G_3 + \widehat{\nu_0^*}\mathscr{Z}G_1 + \widehat{\nu_0^*}\xi G_3\right)\left(\tfrac{1}{K} - \widehat{\nu_0^*}\right)\left(\mathscr{Z}G_1 - \xi G_3\right) \\ &\quad - \tfrac{1}{K}G_1\left(\tfrac{1}{K}G_1 + 2G_3\right) \\ &> \left(\tfrac{1}{K}\mathscr{Z}G_1 + 2G_3 + \widehat{\nu_0^*}\eta G_1\right)\left(\tfrac{1}{K} - \widehat{\nu_0^*}\right)\eta G_1 \\ &\quad - \tfrac{1}{K}G_1\left(\tfrac{1}{K}G_1 + 2G_3\right) \\ &= \left[\eta\left(\tfrac{1}{K}\mathscr{Z}G_1 + 2G_3 + \widehat{\nu_0^*}\eta G_1\right)\left(\tfrac{1}{K} - \widehat{\nu_0^*}\right)\right. \\ &\quad \left. - \tfrac{1}{K}\left(\tfrac{1}{K}G_1 + 2G_3\right)\right]G_1 \\ &= \left\{\left[\left(\tfrac{1}{K}\mathscr{Z} + \widehat{\nu_0^*}\eta\right)\eta G_1 + 2\eta G_3\right]\left(\tfrac{1}{K} - \widehat{\nu_0^*}\right)\right. \\ &\quad \left. - \tfrac{1}{K}\left(\tfrac{1}{K}G_1 + 2G_3\right)\right\}G_1.\end{aligned} \tag{A.252}$$

Now, making use of (A.204), (A.206) and (A.248), we come to

$$\eta = 1 + a_1K > 1 \tag{A.253}$$

and

$$\mathscr{Z} = \eta + a_0\xi > \eta + 2a_0K. \tag{A.254}$$

Then, applying inequalities (A.253) and (A.254) to (A.252) one gets:

$$\begin{aligned}\mathscr{P}_1 &= \left\{\left[\left(\tfrac{1}{K}\mathscr{Z} + \widehat{\nu_0^*}\eta\right)\eta G_1 + 2\eta G_3\right]\left(\tfrac{1}{K} - \widehat{\nu_0^*}\right)\right. \\ &\quad \left. - \tfrac{1}{K}\left(\tfrac{1}{K}G_1 + 2G_3\right)\right\}G_1 \\ &> \left\{\left[\left(\tfrac{1}{K}(\eta + 2a_0K) + \widehat{\nu_0^*}\right)G_1 + 2\eta G_3\right]\left(\tfrac{1}{K} - \widehat{\nu_0^*}\right)\right. \\ &\quad \left. - \tfrac{1}{K}\left(\tfrac{1}{K}G_1 + 2G_3\right)\right\}G_1 \\ &= \left\{\left[\left(\tfrac{1}{K}\eta + 2a_0 + \widehat{\nu_0^*}\right)G_1 + 2\eta G_3\right]\left(\tfrac{1}{K} - \widehat{\nu_0^*}\right)\right. \\ &\quad \left. - \tfrac{1}{K}\left(\tfrac{1}{K}G_1 + 2G_3\right)\right\}G_1.\end{aligned} \tag{A.255}$$

Substituting the value of G_3 given by (A.208) in (A.255), we have:

$$\begin{aligned}\mathscr{P}_1 &> \left\{\left[\left(\tfrac{1}{K}\eta + 2a_0 + \widehat{\nu_0^*}\right)G_1 + 2\eta G_3\right]\left(\tfrac{1}{K} - \widehat{\nu_0^*}\right)\right. \\ &\quad \left. - \tfrac{1}{K}\left(\tfrac{1}{K}G_1 + 2G_3\right)\right\}G_1. \\ &= \left\{\left[\left(\tfrac{1}{K}\eta + 2a_0 + \widehat{\nu_0^*}\right)G_1 + 2\eta\left[a_0G_1 - (b_1 - b_0)\right]\right]\left(\tfrac{1}{K} - \widehat{\nu_0^*}\right)\right. \\ &\quad \left. - \tfrac{1}{K}\left\{\tfrac{1}{K}G_1 + 2\left[a_0G_1 - (b_1 - b_0)\right]\right\}\right\}G_1. \\ &= \left\{\left[\left(\tfrac{1}{K}\eta + 2a_0 + \widehat{\nu_0^*}\right)G_1 + 2a_0\eta G_1 - 2\eta(b_1 - b_0)\right]\left(\tfrac{1}{K} - \widehat{\nu_0^*}\right)\right. \\ &\quad \left. - \tfrac{1}{K}\left[\tfrac{1}{K}G_1 + 2a_0G_1 - 2(b_1 - b_0)\right]\right\}G_1.\end{aligned} \tag{A.256}$$

$$
\begin{aligned}
&= \left\{ \left[\left(\tfrac{1}{K}\eta + 2a_0 + 2a_0\eta + \widehat{\nu_0^*} \right) G_1 - 2\eta(b_1 - b_0) \right] \left(\tfrac{1}{K} - \widehat{\nu_0^*} \right) \right. \\
&\quad \left. - \tfrac{1}{K} \left[\left(\tfrac{1}{K} + 2a_0 \right) G_1 - 2(b_1 - b_0) \right] \right\} G_1. \\
&= \left\{ \left(\tfrac{1}{K}\eta + 2a_0 + 2a_0\eta + \widehat{\nu_0^*} \right) \left(\tfrac{1}{K} - \widehat{\nu_0^*} \right) G_1 - 2\eta \left(\tfrac{1}{K} - \widehat{\nu_0^*} \right) (b_1 - b_0) \right. \\
&\quad \left. - \tfrac{1}{K} \left(\tfrac{1}{K} + 2a_0 \right) G_1 + 2\tfrac{1}{K}(b_1 - b_0) \right\} G_1. \\
&= \left\{ \left[\left(\tfrac{1}{K}\eta + 2a_0 + 2a_0\eta + \widehat{\nu_0^*} \right) \left(\tfrac{1}{K} - \widehat{\nu_0^*} \right) - \tfrac{1}{K} \left(\tfrac{1}{K} + 2a_0 \right) \right] G_1 \right. \\
&\quad \left. + 2 \left[\tfrac{1}{K} - \eta \left(\tfrac{1}{K} - \widehat{\nu_0^*} \right) \right] (b_1 - b_0) \right\} G_1.
\end{aligned}
$$

Next, we substitute the value of η given by (A.204) in (A.256) to obtain:

$$
\begin{aligned}
\mathscr{P}_1 &> \left\{ \left[\left(\tfrac{1}{K}\eta + 2a_0 + 2a_0\eta + \widehat{\nu_0^*} \right) \left(\tfrac{1}{K} - \widehat{\nu_0^*} \right) - \tfrac{1}{K} \left(\tfrac{1}{K} + 2a_0 \right) \right] G_1 \right. \\
&\quad \left. + 2 \left[\tfrac{1}{K} - \eta \left(\tfrac{1}{K} - \widehat{\nu_0^*} \right) \right] (b_1 - b_0) \right\} G_1. \\
&= \left\{ \left[\left(\tfrac{1}{K}(1 + a_1 K) + 2a_0 + 2a_0\eta + \widehat{\nu_0^*} \right) \left(\tfrac{1}{K} - \widehat{\nu_0^*} \right) - \tfrac{1}{K} \left(\tfrac{1}{K} + 2a_0 \right) \right] G_1 \right. \\
&\quad \left. + 2 \left[\tfrac{1}{K} - (1 + a_1 K) \left(\tfrac{1}{K} - \widehat{\nu_0^*} \right) \right] (b_1 - b_0) \right\} G_1. \\
&= \left\{ \left[\left(\tfrac{1}{K} + a_1 + 2a_0 + 2a_0\eta + \widehat{\nu_0^*} \right) \left(\tfrac{1}{K} - \widehat{\nu_0^*} \right) - \tfrac{1}{K} \left(\tfrac{1}{K} + 2a_0 \right) \right] G_1 \right. \\
&\quad \left. + 2 \left[\tfrac{1}{K} - \left(\tfrac{1}{K} - \widehat{\nu_0^*} \right) - a_1 K \left(\tfrac{1}{K} - \widehat{\nu_0^*} \right) \right] (b_1 - b_0) \right\} G_1. \\
&= \left\{ \left[-\widehat{\nu_0^*} \left(\tfrac{1}{K} + 2a_0 \right) + (a_1 + 2a_0\eta) \left(\tfrac{1}{K} - \widehat{\nu_0^*} \right) + \widehat{\nu_0^*} \left(\tfrac{1}{K} - \widehat{\nu_0^*} \right) \right] G_1 \right. \\
&\quad \left. + 2 \left[(1 + a_1 K) \widehat{\nu_0^*} - a_1 \right] (b_1 - b_0) \right\} G_1. \\
&= \left\{ \left[(a_1 + 2a_0\eta) \left(\tfrac{1}{K} - \widehat{\nu_0^*} \right) - \widehat{\nu_0^*} \left(2a_0 + \widehat{\nu_0^*} \right) \right] G_1 \right. \\
&\quad \left. + 2 \left[(1 + a_1 K) \widehat{\nu_0^*} - a_1 \right] (b_1 - b_0) \right\} G_1. \\
&= \left[V_5 G_1 + 2W_5(b_1 - b_0) \right] G_1,
\end{aligned}
\tag{A.257}
$$

where

$$
V_5 = (a_1 + 2a_0\eta) \left(\tfrac{1}{K} - \widehat{\nu_0^*} \right) - \widehat{\nu_0^*} \left(2a_0 + \widehat{\nu_0^*} \right) \tag{A.258}
$$

and

$$
W_5 = (1 + a_1 K) \widehat{\nu_0^*} - a_1. \tag{A.259}
$$

Now, we only need to show that $V_5 > 0$ and $W_5 > 0$.

First, since $\nu_0^*(\beta)$ is strictly decreasing, we have $\widehat{\nu_0^*} = \lim_{\beta \to 0} \nu_0^*(\beta) > \nu_0^*(1) = \overline{\nu_0^*}$, thus,

$$
W_5 = (1 + a_1 K) \widehat{\nu_0^*} - a_1 > (1 + a_1 K) \overline{\nu_0^*} - a_1. \tag{A.260}
$$

Substituting the value of $\overline{\nu_0^*}$ given by (A.193) in (A.260) we have that:

$$
\begin{aligned}
W_5 &> (1 + a_1 K)\,\overline{v_0^*} - a_1 \\
&= (1 + a_1 K)\left(\frac{a_0 + a_1 + a_0 a_1 K}{1 + 2a_0 K + a_1 K + a_0 a_1 K^2}\right) - a_1 \\
&= \frac{(1 + a_1 K)(a_0 + a_1 + a_0 a_1 K) - a_1\left(1 + 2a_0 K + a_1 K + a_0 a_1 K^2\right)}{1 + 2a_0 K + a_1 K + a_0 a_1 K^2} \\
&= \frac{V_6}{W_6},
\end{aligned}
\tag{A.261}
$$

where

$$
V_6 = (1 + a_1 K)(a_0 + a_1 + a_0 a_1 K) - a_1\left(1 + 2a_0 K + a_1 K + a_0 a_1 K^2\right) \tag{A.262}
$$

and

$$
W_6 = 1 + 2a_0 K + a_1 K + a_0 a_1 K^2. \tag{A.263}
$$

Given the values of a_0, a_1 and K, it's easy to see that $W_6 > 0$. Now, we calculate the value of V_6:

$$
\begin{aligned}
V_6 &= (1 + a_1 K)(a_0 + a_1 + a_0 a_1 K) - a_1\left(1 + 2a_0 K + a_1 K + a_0 a_1 K^2\right) \\
&= (a_0 + a_1 + a_0 a_1 K) + a_1 K (a_0 + a_1 + a_0 a_1 K) - a_1\left(1 + 2a_0 K + a_1 K + a_0 a_1 K^2\right) \\
&= a_0 + a_1 (1 + a_0 K) + a_1\left(a_0 K + a_1 K + a_0 a_1 K^2\right) - a_1\left(1 + 2a_0 K + a_1 K + a_0 a_1 K^2\right) \\
&= a_0 + a_1\left(1 + 2a_0 K + a_1 K + a_0 a_1 K^2\right) - a_1\left(1 + 2a_0 K + a_1 K + a_0 a_1 K^2\right) \\
&= a_0 > 0.
\end{aligned}
\tag{A.264}
$$

Therefore $V_6 > 0$ and

$$
W_5 > \frac{V_6}{W_6} > 0. \tag{A.265}
$$

So we only lack showing that $V_5 > 0$:

$$
\begin{aligned}
V_5 &= (a_1 + 2a_0\eta)\left(\tfrac{1}{K} - \widehat{v_0^*}\right) - \widehat{v_0^*}\left(2a_0 + \widehat{v_0^*}\right) \\
&= (a_1 + a_0\eta)\left(\tfrac{1}{K} - \widehat{v_0^*}\right) + a_0\eta\left(\tfrac{1}{K} - \widehat{v_0^*}\right) - \widehat{v_0^*}\left(a_0 + \widehat{v_0^*}\right) - a_0\widehat{v_0^*} \\
&= \left[(a_1 + a_0\eta)\left(\tfrac{1}{K} - \widehat{v_0^*}\right) - \widehat{v_0^*}\left(a_0 + \widehat{v_0^*}\right)\right] + a_0\left[\eta\left(\tfrac{1}{K} - \widehat{v_0^*}\right) - \widehat{v_0^*}\right] \\
&= V_7 + a_0 W_7,
\end{aligned}
\tag{A.266}
$$

where

$$
V_7 = (a_1 + a_0\eta)\left(\tfrac{1}{K} - \widehat{v_0^*}\right) - \widehat{v_0^*}\left(a_0 + \widehat{v_0^*}\right) \tag{A.267}
$$

and

$$
W_7 = \eta\left(\tfrac{1}{K} - \widehat{v_0^*}\right) - \widehat{v_0^*}. \tag{A.268}
$$

Finally, we will demonstrate that $V_7 > 0$ and $W_7 > 0$. Substituting the value of η given by (A.204) in V_7 we get:

$$\begin{aligned}
V_7 &= (a_1 + a_0\eta)\left(\tfrac{1}{K} - \widehat{\nu_0^*}\right) - \widehat{\nu_0^*}\left(a_0 + \widehat{\nu_0^*}\right) \\
&= [a_1 + a_0(1 + a_1 K)]\left(\tfrac{1}{K} - \widehat{\nu_0^*}\right) - \widehat{\nu_0^*}\left(a_0 + \widehat{\nu_0^*}\right) \\
&= (a_1 + a_0 + a_0 a_1 K)\left(\tfrac{1}{K} - \widehat{\nu_0^*}\right) - \widehat{\nu_0^*}\left(a_0 + \widehat{\nu_0^*}\right) \\
&= \tfrac{1}{K}(a_1 + a_0 + a_0 a_1 K) - \widehat{\nu_0^*}(a_1 + a_0 + a_0 a_1 K) - a_0\widehat{\nu_0^*} - \widehat{\nu_0^*}^2 \\
&= \tfrac{1}{K}(a_1 + a_0 + a_0 a_1 K) - \widehat{\nu_0^*}(a_1 + 2a_0 + a_0 a_1 K) - \widehat{\nu_0^*}^2.
\end{aligned} \tag{A.269}$$

we get:

$$(1-\beta)\left(-2\tau + a_1\tau^2\right)\nu_0^2 + \left(\beta - 2a_0\tau - \beta a_1\tau + a_0 a_1\tau^2\right)\nu_0 - (a_0 + a_1 - a_0 a_1\tau) = 0,$$

given by (A.16), for $\tau = -K$. Now by applying the limit when $\beta \to 0$, one obtains the following equality:

$$\left(2K + a_1K^2\right)\widehat{\nu_0^*}^2 + \left(2a_0K + a_0a_1K^2\right)\widehat{\nu_0^*} - (a_0 + a_1 + a_0a_1K) = 0. \tag{A.270}$$

Therefore,

$$\begin{aligned}
&\left(2K + a_1K^2\right)\widehat{\nu_0^*}^2 + \left(2a_0K + a_0a_1K^2\right)\widehat{\nu_0^*} - (a_0 + a_1 + a_0a_1K) \\
&= (2 + a_1K)K\widehat{\nu_0^*}^2 + (2a_0 + a_0a_1K)K\widehat{\nu_0^*} - \tfrac{1}{K}(a_0 + a_1 + a_0a_1K)K \\
&= \left[(2 + a_1K)\widehat{\nu_0^*}^2 + (2a_0 + a_0a_1K)\widehat{\nu_0^*} - \tfrac{1}{K}(a_0 + a_1 + a_0a_1K)\right]K \\
&= \left[(1 + 1 + a_1K)\widehat{\nu_0^*}^2 + (a_1 + 2a_0 + a_0a_1K - a_1)\widehat{\nu_0^*} - \tfrac{1}{K}(a_0 + a_1 + a_0a_1K)\right]K \\
&= \left[\widehat{\nu_0^*}^2 + (a_1 + 2a_0 + a_0a_1K)\widehat{\nu_0^*} - \tfrac{1}{K}(a_0 + a_1 + a_0a_1K) + (1 + a_1K)\widehat{\nu_0^*}^2 - a_1\widehat{\nu_0^*}\right]K \\
&= 0.
\end{aligned} \tag{A.271}$$

Since $K > 0$, we have that

$$\widehat{\nu_0^*}^2 + (a_1 + 2a_0 + a_0a_1K)\widehat{\nu_0^*} - \tfrac{1}{K}(a_0 + a_1 + a_0a_1K) + (1 + a_1K)\widehat{\nu_0^*}^2 - a_1\widehat{\nu_0^*} = 0, \tag{A.272}$$

thus,

$$(1 + a_1K)\widehat{\nu_0^*}^2 - a_1\widehat{\nu_0^*} = \tfrac{1}{K}(a_0 + a_1 + a_0a_1K) - (a_1 + 2a_0 + a_0a_1K)\widehat{\nu_0^*} - \widehat{\nu_0^*}^2. \tag{A.273}$$

Applying equality (A.273) to (A.269) we have that:

$$
\begin{aligned}
V_7 =& \tfrac{1}{K}\left(a_0 + a_1 + a_0 a_1 K\right) - \left(a_1 + 2a_0 + a_0 a_1 K\right)\widehat{v_0^*} - \widehat{v_0^*}^2 \\
=& \left(1 + a_1 K\right)\widehat{v_0^*}^2 - a_1 \widehat{v_0^*} \\
=& \left[\left(1 + a_1 K\right)\widehat{v_0^*} - a_1\right]\widehat{v_0^*} \\
=& W_5 \widehat{v_0^*}.
\end{aligned}
\tag{A.274}
$$

Now recalling that $W_5 > 0$ and $\widehat{v_0^*} > 0$, we have:

$$
V_7 = W_5 \widehat{v_0^*} > 0. \tag{A.275}
$$

So we only need to show that $W_7 > 0$. To do this, we plug the value of η given by (A.204) in W_7 to get:

$$
\begin{aligned}
W_7 =& \eta\left(\tfrac{1}{K} - \widehat{v_0^*}\right) - \widehat{v_0^*} \\
=& \left(1 + a_1 K\right)\left(\tfrac{1}{K} - \widehat{v_0^*}\right) - \widehat{v_0^*} \\
=& \tfrac{1}{K}\left(1 + a_1 K\right) - \left(1 + a_1 K\right)\widehat{v_0^*} - \widehat{v_0^*} \\
=& \tfrac{1}{K}\left(1 + a_1 K\right) - \left(2 + a_1 K\right)\widehat{v_0^*}.
\end{aligned}
\tag{A.276}
$$

Using relationship (A.270) we have that:

$$
\begin{aligned}
&\left(2K + a_1 K^2\right)\widehat{v_0^*}^2 + \left(2a_0 K + a_0 a_1 K^2\right)\widehat{v_0^*} - \left(a_0 + a_1 + a_0 a_1 K\right) \\
&= \left[\left(2 + a_1 K\right)\widehat{v_0^*}^2 + \left(2a_0 + a_0 a_1 K\right)\widehat{v_0^*} - \tfrac{1}{K}\left(a_0 + a_1 + a_0 a_1 K\right)\right] K \\
&= \left\{\left(2 + a_1 K\right)\widehat{v_0^*}^2 + a_0\left(2 + a_1 K\right)\widehat{v_0^*} - \tfrac{1}{K}\left[a_1 + a_0\left(1 + a_1 K\right)\right]\right\} K \\
&= \left\{\left(2 + a_1 K\right)\widehat{v_0^*}^2 + a_0\left(2 + a_1 K\right)\widehat{v_0^*} - \tfrac{1}{K}a_1 - \tfrac{1}{K}a_0\left(1 + a_1 K\right)\right\} K \\
&= \left\{\left(2 + a_1 K\right)\widehat{v_0^*}^2 - \tfrac{1}{K}a_1 + a_0\left[\left(2 + a_1 K\right)\widehat{v_0^*} - \tfrac{1}{K}\left(1 + a_1 K\right)\right]\right\} K.
\end{aligned}
\tag{A.277}
$$

Since $K > 0$, we have that:

$$
\left(2 + a_1 K\right)\widehat{v_0^*}^2 - \tfrac{1}{K}a_1 + a_0\left[\left(2 + a_1 K\right)\widehat{v_0^*} - \tfrac{1}{K}\left(1 + a_1 K\right)\right] = 0, \tag{A.278}
$$

which implies

$$
\left(2 + a_1 K\right)\widehat{v_0^*}^2 - \tfrac{1}{K}a_1 = -a_0\left[\left(2 + a_1 K\right)\widehat{v_0^*} - \tfrac{1}{K}\left(1 + a_1 K\right)\right]. \tag{A.279}
$$

As $a_0 > 0$, one gets:

$$
\tfrac{1}{a_0}\left[\left(2 + a_1 K\right)\widehat{v_0^*}^2 - \tfrac{1}{K}a_1\right] = \tfrac{1}{K}\left(1 + a_1 K\right) - \left(2 + a_1 K\right)\widehat{v_0^*}. \tag{A.280}
$$

Now, applying equality (A.280) to (A.276), we deduce:

$$\begin{aligned} W_7 =& \tfrac{1}{K}\left(1 + a_1 K\right) - \left(2 + a_1 K\right) \widehat{\nu_0^*} \\ =& \tfrac{1}{a_0}\left[\left(2 + a_1 K\right) \widehat{\nu_0^*}^2 - \tfrac{1}{K} a_1\right] \\ =& \frac{U_8}{a_0}, \end{aligned} \tag{A.281}$$

where

$$U_8 = \left(2 + a_1 K\right) \widehat{\nu_0^*}^2 - \tfrac{1}{K} a_1. \tag{A.282}$$

Finally, let's suppose that

$$U_8 \leq 0. \tag{A.283}$$

Substituting the value of $\widehat{\nu_0^*}$, given by (A.141), in U_8 we have:

$$\begin{aligned} U_8 =& \left(2 + a_1 K\right) \widehat{\nu_0^*}^2 - \tfrac{1}{K} a_1 \\ =& \left(2 + a_1 K\right) \left[\frac{2(a_0 + a_1 + a_0 a_1 K)}{(2a_0 K + a_0 a_1 K^2) + \sqrt{(2a_0 K + a_0 a_1 K^2)^2 + 4(2K + a_1 K^2)(a_0 + a_1 + a_0 a_1 K)}}\right]^2 - \tfrac{1}{K} a_1 \\ =& \Bigg\{\left(2 + a_1 K\right)\left[2\left(a_0 + a_1 + a_0 a_1 K\right)\right]^2 - \tfrac{1}{K} a_1 \Bigg[\left(2a_0 K + a_0 a_1 K^2\right) \\ & + \sqrt{\left(2a_0 K + a_0 a_1 K^2\right)^2 + 4\left(2K + a_1 K^2\right)\left(a_0 + a_1 + a_0 a_1 K\right)}\Bigg]^2\Bigg\} \\ & / \left[\left(2a_0 K + a_0 a_1 K^2\right) + \sqrt{\left(2a_0 K + a_0 a_1 K^2\right)^2 + 4\left(2K + a_1 K^2\right)\left(a_0 + a_1 + a_0 a_1 K\right)}\right]^2 \\ =& \frac{V_9}{W_9} \leq 0, \end{aligned} \tag{A.284}$$

where

$$\begin{aligned} V_9 =& \left(2 + a_1 K\right)\left[2\left(a_0 + a_1 + a_0 a_1 K\right)\right]^2 - \tfrac{1}{K} a_1 \Bigg[\left(2a_0 K + a_0 a_1 K^2\right) \\ & + \sqrt{\left(2a_0 K + a_0 a_1 K^2\right)^2 + 4\left(2K + a_1 K^2\right)\left(a_0 + a_1 + a_0 a_1 K\right)}\Bigg]^2 \end{aligned} \tag{A.285}$$

and

$$W_9 = \left[\left(2a_0 K + a_0 a_1 K^2\right) + \sqrt{\left(2a_0 K + a_0 a_1 K^2\right)^2 + 4\left(2K + a_1 K^2\right)\left(a_0 + a_1 + a_0 a_1 K\right)}\right]^2. \tag{A.286}$$

For any given values of a_0, a_1 and K, one has $W_9 > 0$. Therefore, Eq. (A.284) implies

$$V_9 \leq 0. \tag{A.287}$$

Hence

$$\begin{aligned}
V_9 =& (2 + a_1 K)\,[2\,(a_0 + a_1 + a_0 a_1 K)]^2 - \tfrac{1}{K} a_1 \Big[\Big(2a_0 K + a_0 a_1 K^2\Big) \\
&+ \sqrt{\left(2a_0 K + a_0 a_1 K^2\right)^2 + 4\,(2K + a_1 K^2)\,(a_0 + a_1 + a_0 a_1 K)}\Big]^2 \\
=& 4\,(2 + a_1 K)\,[a_1 + a_0\,(1 + a_1 K)]^2 - \tfrac{1}{K} a_1 \Big[a_0 K\,(2 + a_1 K) \\
&+ \sqrt{K\,(2 + a_1 K)\,\{a_0^2 K\,(2 + a_1 K) + 4\,[a_1 + a_0\,(1 + a_1 K)]\}}\Big]^2 \\
=& 4a_1^2\,(2 + a_1 K) + 8a_0 a_1\,(2 + a_1 K)\,(1 + a_1 K) + 4a_0^2\,(2 + a_1 K)\,(1 + a_1 K)^2 \\
&- a_1 a_0^2 K\,(2 + a_1 K)^2 - a_1\,(2 + a_1 K)\Big\{a_0^2 K\,(2 + a_1 K) + 4\,[a_1 + a_0\,(1 + a_1 K)]\Big\} \\
&- 2a_0 a_1\,(2 + a_1 K)\sqrt{K\,(2 + a_1 K)\,\{a_0^2 K\,(2 + a_1 K) + 4\,[a_1 + a_0\,(1 + a_1 K)]\}} \\
=& (2 + a_1 K)\Big[4a_1^2 + 8a_0 a_1\,(1 + a_1 K) + 4a_0^2\,(1 + a_1 K)^2 \\
&- a_1 a_0^2 K\,(2 + a_1 K) - a_1 a_0^2 K\,(2 + a_1 K) - 4a_1\,[a_1 + a_0\,(1 + a_1 K)]\Big] \\
&- 2a_0 a_1\,(2 + a_1 K)\sqrt{K\,(2 + a_1 K)\,\{a_0^2 K\,(2 + a_1 K) + 4\,[a_1 + a_0\,(1 + a_1 K)]\}} \\
=& (2 + a_1 K)\Big[4a_0 a_1\,(1 + a_1 K) + 4a_0^2\,(1 + a_1 K)^2 - 2a_1 a_0^2 K\,(2 + a_1 K)\Big] \\
&- 2a_0 a_1\,(2 + a_1 K)\sqrt{K\,(2 + a_1 K)\,\{a_0^2 K\,(2 + a_1 K) + 4\,[a_1 + a_0\,(1 + a_1 K)]\}} \\
=& a_0\,(2 + a_1 K)\Big[4a_1\,(1 + a_1 K) + 4a_0\Big(1 + 2a_1 K + a_1^2 K^2\Big) - 2a_0 a_1 K\,(2 + a_1 K)\Big] \\
&- 2a_0 a_1\,(2 + a_1 K)\sqrt{K\,(2 + a_1 K)\,\{a_0^2 K\,(2 + a_1 K) + 4\,[a_1 + a_0\,(1 + a_1 K)]\}} \\
=& a_0\,(2 + a_1 K)\Big(4a_0 + 4a_1 + 4a_0 a_1 K + 4a_1^2 K + 2a_0 a_1^2 K^2\Big) \\
&- 2a_0 a_1\,(2 + a_1 K)\sqrt{K\,(2 + a_1 K)\,\{a_0^2 K\,(2 + a_1 K) + 4\,[a_1 + a_0\,(1 + a_1 K)]\}} \\
=& 2a_0\,(2 + a_1 K)\Big[\Big(2a_0 + 2a_1 + 2a_0 a_1 K + 2a_1^2 K + a_0 a_1^2 K^2\Big) \\
&- a_1\sqrt{K\,(2 + a_1 K)\,\{a_0^2 K\,(2 + a_1 K) + 4\,[a_1 + a_0\,(1 + a_1 K)]\}}\Big] \\
=& 2a_0\,(2 + a_1 K)\Big[2a_0 + a_1 K\,(2\tfrac{1}{K} + 2a_0 + 2a_1 + a_0 a_1 K) \\
&- a_1\sqrt{K\,(2 + a_1 K)\,\{a_0^2 K\,(2 + a_1 K) + 4\,[a_1 + a_0\,(1 + a_1 K)]\}}\Big] \leq 0.
\end{aligned} \tag{A.288}$$

Since $2a_0\,(2 + a_1 K) > 0$, then, from (A.288) the following condition must be met:

$$\begin{aligned}&2a_0 + a_1 K \left(2\tfrac{1}{K} + 2a_0 + 2a_1 + a_0 a_1 K\right)\\&- a_1\sqrt{K\left(2 + a_1 K\right)\left\{a_0^2 K\left(2 + a_1 K\right) + 4\left[a_1 + a_0\left(1 + a_1 K\right)\right]\right\}} \le 0.\end{aligned} \tag{A.289}$$

Therefore,

$$\begin{aligned}&2a_0 + a_1 K \left(2\tfrac{1}{K} + 2a_0 + 2a_1 + a_0 a_1 K\right)\\&\le a_1\sqrt{K\left(2 + a_1 K\right)\left\{a_0^2 K\left(2 + a_1 K\right) + 4\left[a_1 + a_0\left(1 + a_1 K\right)\right]\right\}}.\end{aligned} \tag{A.290}$$

Since both sides of inequality (A.290) are positive, then,

$$\begin{aligned}&\left[2a_0 + a_1 K \left(2\tfrac{1}{K} + 2a_0 + 2a_1 + a_0 a_1 K\right)\right]^2\\&\le a_1^2\left[K\left(2 + a_1 K\right)\left\{a_0^2 K\left(2 + a_1 K\right) + 4\left[a_1 + a_0\left(1 + a_1 K\right)\right]\right\}\right],\end{aligned} \tag{A.291}$$

which leads to

$$\begin{aligned}0 \le\, & a_1^2\left[K\left(2 + a_1 K\right)\left\{a_0^2 K\left(2 + a_1 K\right) + 4\left[a_1 + a_0\left(1 + a_1 K\right)\right]\right\}\right]\\
&- \left[2a_0 + a_1 K\left(2\tfrac{1}{K} + 2a_0 + 2a_1 + a_0 a_1 K\right)\right]^2\\
=\,& a_1^2 K\left(2 + a_1 K\right)\left(2a_0^2 K + a_0^2 a_1 K^2 + 4a_1 + 4a_0 + 4a_0 a_1 K\right)\\
&- \left(2a_0 + 2a_1 + 2a_0 a_1 K + 2a_1^2 K + a_0 a_1^2 K^2\right)^2\\
=\,& a_1^2 K\left(2 + a_1 K\right)\left[4\left(a_0 + a_1 + a_0 a_1 K\right) + a_0^2 K\left(2 + a_1 K\right)\right]\\
&- \left[2\left(a_0 + a_1 + a_0 a_1 K\right) + a_1^2 K\left(2 + a_0 K\right)\right]^2\\
=\,& 4a_1^2 K\left(a_0 + a_1 + a_0 a_1 K\right)\left[\left(2 + a_1 K\right) - \left(2 + a_0 K\right)\right]\\
&+ a_1^2 K^2\left[a_0^2\left(2 + a_1 K\right)^2 - a_1^2\left(2 + a_0 K\right)^2\right]\\
&- 4\left(a_0 + a_1 + a_0 a_1 K\right)^2\\
=\,& 4a_1^2 K^2\left(a_0 + a_1 + a_0 a_1 K\right)\left(a_1 - a_0\right)\\
&+ a_1^2 K^2\left(2a_0 + a_0 a_1 K + 2a_1 + a_0 a_1 K\right)\left(2a_0 + a_0 a_1 K - 2a_1 - a_0 a_1 K\right)\\
&- 4\left(a_0 + a_1 + a_0 a_1 K\right)^2\\
=\,& 4a_1^2 K^2\left(a_0 + a_1 + a_0 a_1 K\right)\left(a_1 - a_0\right)\\
&- 4a_1^2 K^2\left(a_0 + a_1 + a_0 a_1 K\right)\left(a_1 - a_0\right)\\
&- 4\left(a_0 + a_1 + a_0 a_1 K\right)^2\\
=\,& - 4\left(a_0 + a_1 + a_0 a_1 K\right)^2.\end{aligned} \tag{A.292}$$

Thus,

$$-4\left(a_0 + a_1 + a_0 a_1 K\right)^2 \ge 0,$$

which cannot happen due to $a_0 + a_1 + a_0 a_1 K > 0$.

Therefore, the assumption was false, so $U_8 > 0$. Thus,

$$W_7 = \frac{U_8}{a_0} > 0, \tag{A.293}$$

then,

$$V_5 = V_7 + a_0 W_7 > 0, \tag{A.294}$$

which proves that

$$\mathscr{P}_1 > [V_5 G_1 + 2W_5(b_1 - b_0)]\, G_1 \geq V_5 {G_1}^2 > 0. \tag{A.295}$$

So we have that $\mathscr{P}_1 > 0$, $\mathscr{Q}_1 > 0$ and $\mathscr{R}_1 > 0$, then,

$$V_2 = \mathscr{P}_1 + \mathscr{Q}_1 + \mathscr{R}_1 > 0, \tag{A.296}$$

and therefore,

$$\lim_{\beta \to 0} (\pi_1^c - \pi_1^*)(\beta) = \frac{1}{2} \frac{V_2}{W_2} > 0, \tag{A.297}$$

which proves (A.192). The proof of the theorem is complete ■

Appendix B: Proofs of Results from Chapter 3

B.1 Proofs of Results from Sect. 3.3

Theorem 3.2 *Let the number of oligopoly producers be at least three, i.e.,* $n \geq 3$, *then, under assumptions* **A4–A6**, *there exists an interior equilibrium. Moreover, if the number of producers is two, i.e.,* $n = 2$, *in addition to assumptions* **A4–A6**, *suppose that there exists an* $\varepsilon > 0$ *such that* $G'(p) \leq -\varepsilon$ *for all* $p > 0$, *then, there exists interior equilibrium.*

Proof For any given set of influence coefficients $\nu = (\nu_1, \dots, \nu_n) \geq 0$, by Theorem 3.1, there exists the unique exterior equilibrium $(p(\nu), q_1(\nu), \dots, q_n(\nu))$.

Now, we define the following functions:

$$F_i(\nu) = \frac{1}{\displaystyle\sum_{\substack{j=1 \\ j \neq i}}^{n} \frac{1}{\nu_j + f_j''(q_j(\nu))} - G'(p(\nu))}, \quad i = 1, \dots, n. \tag{B.1}$$

These functions are well-defined and continuous with respect to $\nu = (\nu_1, \dots, \nu_n) \geq 0$, due to assumptions **A**4 and **A**5.

Therefore, the function $H = (F_1, \dots, F_n) : \mathbb{R}^n_+ \to \mathbb{R}^n_+$ is also continuous.

Next, we define the value $s = \max\{f_i''(q_i) \mid 0 \leq q_i \leq G(p_0),\ i = 1, \dots, n\} > 0$.

For $n \geq 3$, if $0 \leq \nu_i \leq \dfrac{s}{n-2}$ for all $i = 1, \dots, n$, we have:

J. G. Flores Muñiz et al., *Public Interest and Private Enterprize: New Developments*, Lecture Notes in Networks and Systems 138,
https://doi.org/10.1007/978-3-030-58349-1

$$0 \leq F_i(\nu) = \frac{1}{\sum\limits_{\substack{j=1 \\ j \neq i}}^{n} \frac{1}{\nu_j + f_j''(q_j(\nu))} - G'(p(\nu))} \leq \frac{1}{\sum\limits_{\substack{j=1 \\ j \neq i}}^{n} \frac{1}{\nu_j + f_j''(q_j(\nu))}}$$
$$\leq \frac{1}{\sum\limits_{\substack{j=1 \\ j \neq i}}^{n} \frac{1}{\frac{s}{n-2} + s}} = \frac{1}{\frac{n-1}{\frac{s}{n-2} + s}} = \frac{s}{n-2}, \quad i = 1, \ldots, n. \tag{B.2}$$

Thus, the function $H = (F_1, \ldots, F_n)$ maps the convex compact subset $\left[0, \frac{s}{n-2}\right]^n$ onto itself. Therefore, by Brouwer's fixed-point theorem, H has a fixed point, i.e., there exists $\nu^* = (\nu_1^*, \ldots, \nu_n^*) \geq 0$ such that $F_i(\nu^*) = \nu_i^*$ for all $i = 1, \ldots, n$.

On the other hand, for $n = 2$ and $G'(p) \leq -\varepsilon$ for some $\varepsilon > 0$, if $0 \leq \nu_i \leq \frac{1}{\varepsilon}$ for all $i = 1, \ldots, n$, we have:

$$0 \leq F_1(\nu) = \frac{1}{\frac{1}{\nu_2 + f_2''(q_2(\nu))} - G'(p(\nu))} \leq \frac{1}{-G'(p(\nu))} \leq \frac{1}{\varepsilon},$$
$$0 \leq F_2(\nu) = \frac{1}{\frac{1}{\nu_1 + f_1''(q_1(\nu))} - G'(p(\nu))} \leq \frac{1}{-G'(p(\nu))} \leq \frac{1}{\varepsilon}. \tag{B.3}$$

Thus, the function $H = (F_1, F_2)$ maps the convex compact subset $\left[0, \frac{1}{\varepsilon}\right]^2$ onto itself, then, again by Brouwer's fixed-point theorem, H has a fixed point $F_i(\nu^*) = \nu_i^*$, $i = 1, 2$.

By the definition of the functions (B.1), the influence coefficients $\nu^* = (\nu_1^*, \ldots, \nu_n^*) \geq 0$, given by Brouwer's fixed-point theorem, satisfy the Consistency Criterion 3.2 and, therefore, the vector $(p(\nu^*), q_1(\nu^*), \ldots, q_n(\nu^*), \nu_1^*, \ldots, \nu_n^*)$ is the interior equilibrium. The proof is complete ■

B.2 Proofs of Results from Sect. 3.4

Theorem 3.6 *Suppose that the stronger assumption* **A7** *is true, together with* **A4** *and* **A6**, *and suppose that the function G is concave. Then, the Consistency Criterion for the original oligopoly is a necessary and sufficient condition for the collection of influence conjectures $\nu = (\nu_1, \ldots, \nu_n)$ to produce Nash equilibrium in the meta-game.*

Proof Note that the necessity is a particular case of Theorem 3.5, thus, to prove Theorem 3.6, we just need to establish the sufficiency.

We assume **A**7, that is, for all i, the cost functions f_i are quadratic (and strictly convex) with $f_i(0) = 0$, $f_i'(0) > 0$, and $f_i''(0) > 0$, i.e.,

$$f_i(q_i) = \frac{1}{2} a_i {q_i}^2 + b_i q_i,$$

where $a_i > 0, b_i > 0, i = 1, \dots, n$. Now we are in a position to demonstrate that in this specific case, each interior equilibrium $(p^*; q_1^*, \dots, q_n^*; \nu_1^*, \dots, \nu_n^*)$ of the original oligopoly provides the Nash equilibrium in the meta-game $\Gamma = (N, V, \Pi, D)$. Namely, the consistent conjectures (influence coefficients) $\nu^* = (\nu_1^*, \dots, \nu_n^*)$ satisfying (3.14) form the Nash equilibrium in the meta-game.

Indeed, first of all, Eq. (3.14) in this particular case are reduced to the system

$$\nu_i^* = \frac{1}{\sum\limits_{\substack{k=1\\k\neq i}}^{n} \frac{1}{\nu_k^* + a_k} - G'(p^*)}, \quad i = 1, \dots, n, \tag{B.4}$$

which clearly implies that all components of the vector ν^* are positive: $\nu_i^* > 0$, $i = 1, \dots, n$.

Next, equations

$$\begin{aligned}
\frac{\partial \pi_i}{\partial \nu_i} &= \frac{q_i^2}{\nu_i + f_i''(q_i)} \left[\frac{1}{\sum\limits_{k=1}^{n} \frac{1}{\nu_k + f_k''(q_k)} - G'(p)} - \frac{\sum\limits_{\substack{k=1\\k\neq i}}^{n} \frac{1}{\nu_k + f_k''(q_k)} - G'(p)}{\sum\limits_{k=1}^{n} \frac{1}{\nu_k + f_k''(q_k)} - G'(p)} \nu_i \right] \\
&= \frac{q_i^2}{\nu_i + f_i''(q_i)} \frac{\sum\limits_{\substack{k=1\\k\neq i}}^{n} \frac{1}{\nu_k + f_k''(q_k)} - G'(p)}{\sum\limits_{k=1}^{n} \frac{1}{\nu_k + f_k''(q_k)} - G'(p)} \left[\frac{1}{\sum\limits_{\substack{k=1\\k\neq i}}^{n} \frac{1}{\nu_k + f_k''(q_k)} - G'(p)} - \nu_i \right] \\
&= 0, \quad i = 1, \dots, n.
\end{aligned} \tag{B.5}$$

guarantee that the first-order optimality conditions for the meta-game payoff functions hold:

$$\frac{\partial \pi_i}{\partial \nu_i}(\nu^*) = 0, \quad i = 1, \dots, n. \tag{B.6}$$

Therefore, the value ν_i^* may be the maximum point of the i-th producer's payoff function

$$\tilde{\pi}_i(\nu_i) \equiv \pi(\nu_i, \nu_{-i}^*), \quad i = 1, \dots, n, \tag{B.7}$$

where $\nu_{-i}^* = (\nu_1^*, \dots, \nu_{i-1}^*, \nu_{i+1}^*, \dots, \nu_n^*)$. In order to establish the maximum point property, we are going to fix an arbitrary i and to show that the function $\tilde{\pi}_i = \tilde{\pi}_i(\nu_i)$:

(a) doesn't increase along the ray $(\nu_i^*, +\infty)$,
(b) doesn't decrease in the interval $(0, \nu_i^*)$.

In order to prove (a), taking into account (B.5), it suffices to show that

$$\frac{1}{\sum\limits_{\substack{k=1\\k\neq i}}^{n} \frac{1}{\nu_k^* + a_k} - G'(p(\nu_i^* + \delta, \nu_{-i}^*))} - (\nu_i^* + \delta) \leq 0, \quad \forall \delta > 0. \tag{B.8}$$

By inverting both sides of the consistency equation (B.4) one gets

$$\frac{1}{\nu_i^*} = \sum_{\substack{k=1\\k\neq i}}^{n} \frac{1}{\nu_k^* + a_k} - G'(p^*), \tag{B.9}$$

which clearly implies the relationships

$$\sum_{\substack{k=1\\k\neq i}}^{n} \frac{1}{\nu_k^* + a_k} = \frac{1}{\nu_i^*} + G'(p^*) = \frac{1 + \nu_i^* G'(p^*)}{\nu_i^*} > 0. \tag{B.10}$$

Making use of (B.10), rewrite the left-hand side of (B.8) in the form

$$\begin{aligned}
&\frac{1}{\sum\limits_{\substack{k=1\\k\neq i}}^{n} \frac{1}{\nu_k^* + a_k} - G'(p(\nu_i^* + \delta, \nu_{-i}^*))} - (\nu_i^* + \delta) \\
&= \frac{1}{\frac{1}{\nu_i^*} + G'(p^*) - G'(p(\nu_i^* + \delta, \nu_{-i}^*))} - (\nu_i^* + \delta) \\
&= \frac{\nu_i^*}{1 + \nu_i^* G'(p^*) - \nu_i^* G'(p(\nu_i^* + \delta, \nu_{-i}^*))} - (\nu_i^* + \delta) \\
&= \frac{(\nu_i^*)^2 \left[G'(p(\nu_i^* + \delta, \nu_{-i}^*)) - G'(p^*)\right] - \delta + \nu_i^* \delta \left[G'(p(\nu_i^* + \delta, \nu_{-i}^*)) - G'(p^*)\right]}{1 + \nu_i^* G'(p^*) - \nu_i^* G'(p(\nu_i^* + \delta, \nu_{-i}^*))} \\
&= \frac{\nu_i^* \left[G'(p(\nu_i^* + \delta, \nu_{-i}^*)) - G'(p^*)\right](\nu_i^* + \delta) - \delta}{1 + \nu_i^* G'(p^*) - \nu_i^* G'(p(\nu_i^* + \delta, \nu_{-i}^*))}.
\end{aligned} \tag{B.11}$$

Since $1 + \nu_i^* G'(p^*) > 0$ from (B.10), and $-\nu_i^* G'(p(\nu_i^* + \delta, \nu_{-i}^*)) \geq 0$ by assumption **A**4, then the denominator of (B.11) is strictly positive, thus the sign of ratio (B.11) is determined by that of its numerator. Now since the derivative $G'(p)$ is non-increasing by hypothesis, and $\frac{\partial p}{\partial \nu_i} > 0$ by (3.10), it isn't difficult to show that $\left[G'(p(\nu_i^* + \delta, \nu_{-i}^*)) - G'(p^*)\right] \leq 0$, hence the numerator of (B.11) is strictly negative for any $\delta > 0$:

$$\nu_i^* \left[G'(p(\nu_i^* + \delta, \nu_{-i}^*)) - G'(p^*)\right] (\nu_i^* + \delta) - \delta < 0, \quad \forall \delta > 0. \tag{B.12}$$

The latter brings about the desired inequality

$$\frac{d\tilde{\pi}_i}{d\nu_i}(\nu_i, \nu_{-i}^*) < 0, \quad \forall \nu_i > \nu_i^*, \tag{B.13}$$

which finishes the proof of (a).

Now to demonstrate that (b) is also true, again taking into account (B.5), it is enough to check that

$$\frac{1}{\sum\limits_{\substack{k=1\\k\neq i}}^{n} \frac{1}{\nu_k^* + a_k} - G'(p(\nu_i^* - \delta, \nu_{-i}^*))} - (\nu_i^* - \delta) \geq 0, \quad \forall \delta \text{ such that } 0 < \delta < \nu_i^*. \tag{B.14}$$

Again employing (B.10) yields the following transformation of the left-hand side of (B.14):

$$\begin{aligned}
&\frac{1}{\sum\limits_{\substack{k=1\\k\neq i}}^{n} \frac{1}{\nu_k^* + a_k} - G'(p(\nu_i^* - \delta, \nu_{-i}^*))} - (\nu_i^* - \delta) \\
&\quad = \frac{1}{\frac{1}{\nu_i^*} + G'(p^*) - G'(p(\nu_i^* - \delta, \nu_{-i}^*))} - (\nu_i^* - \delta) \\
&\quad = \frac{\nu_i^*}{1 + \nu_i^* G'(p^*) - \nu_i^* G'(p(\nu_i^* - \delta, \nu_{-i}^*))} - (\nu_i^* - \delta) \\
&\quad = \frac{(\nu_i^*)^2 \left[G'(p(\nu_i^* - \delta, \nu_{-i}^*)) - G'(p^*)\right] + \delta - \nu_i^* \delta \left[G'(p(\nu_i^* - \delta, \nu_{-i}^*)) - G'(p^*)\right]}{1 + \nu_i^* G'(p^*) - \nu_i^* G'(p(\nu_i^* - \delta, \nu_{-i}^*))} \\
&\quad = \frac{\nu_i^* \left[G'(p(\nu_i^* - \delta, \nu_{-i}^*)) - G'(p^*)\right] (\nu_i^* - \delta) + \delta}{1 + \nu_i^* G'(p^*) - \nu_i^* G'(p(\nu_i^* - \delta, \nu_{-i}^*))}.
\end{aligned} \tag{B.15}$$

Similar to the proof of case (a), the denominator of the fraction (B.15) is strictly positive, hence, the fraction's sign coincides with that of its numerator. Again, since the derivative $G'(p)$ is non-increasing by hypothesis, and $\dfrac{\partial p}{\partial \nu_i} > 0$ by (3.10), it is evident that $\left[G'(p(\nu_i^* - \delta, \nu_{-i}^*)) - G'(p^*)\right] \geq 0$, hence, the numerator of (B.15) is strictly positive for any $0 < \delta < \nu_i^*$:

$$\nu_i^* \left[G'(p(\nu_i^* - \delta, \nu_{-i}^*)) - G'(p^*)\right](\nu_i^* - \delta) + \delta > 0, \quad \forall \delta \text{ that } 0 < \delta < \nu_i^*, \tag{B.16}$$

which deduces the desired inequality:

$$\frac{d\tilde{\pi}_i}{d\nu_i}(\nu_i, \nu_{-i}^*) > 0, \quad \forall \nu_i < \nu_i^*. \tag{B.17}$$

Therefore, the proof of (b) is also completed.

Now we can conclude that the Nash equilibrium condition has been established:

$$\pi_i(\nu^*) = \max_{\nu_i > 0} \pi_i(\nu_i, \nu_{-i}^*), \text{ for any } i \in \{1, \ldots, n\}, \tag{B.18}$$

which finishes the proof of Theorem 3.6 ■

Theorem 3.8 *Suppose that apart from assumptions A4, A6 and A7, the regular demand function's derivative is Lipschitz continuous. In more detail, for $n \geq 3$ assume that for any $p_1 > 0$ and $p_2 > 0$ the following inequality holds:*

$$|G'(p_1) - G'(p_2)| \leq \frac{1}{2s^2 G(p_0)} |p_1 - p_2|, \tag{B.19}$$

where $s = \max\{a_1, \ldots, a_n\}$, and the price p_0 is the one defined in the assumption A6. Next, if $n = 2$ (duopoly), we again suppose that there exists $\varepsilon > 0$ such that $G'(p) \leq -\varepsilon$ for all $p > 0$, and the Lipschitz continuity of the demand function is described in the form:

$$|G'(p_1) - G'(p_2)| \leq \frac{2}{\left(\dfrac{a_1 + a_2}{\varepsilon \min\{a_1, a_2\}} + 3\max\{a_1, a_2\}\right)^2 G(p_0)} |p_1 - p_2|, \ \forall p_1, p_2 > 0. \tag{B.20}$$

Then, the Consistency Criterion for the original oligopoly is a necessary and sufficient condition for the collection of influence conjectures $\nu = (\nu_1, \ldots, \nu_n)$ to be the Nash equilibrium in the meta-game.

Proof Again, the necessity is just a particular case of Theorem 3.5, then, we proceed to show the sufficiency.

Just like in the proof of Theorem 3.6, we need to establish that the i-th producer's payoff function

$$\tilde{\pi}_i(\nu_i) \equiv \pi(\nu_i, \nu_{-i}^*), \quad i = 1, \ldots, n, \tag{B.7}$$

has a maximum point at $\nu_i = \nu_i^*$ for a fixed value of i, for which, again, it will suffice to show that:

(a)

$$\nu_i^* \left[G'(p(\nu_i^* + \delta, \nu_{-i}^*)) - G'(p^*)\right](\nu_i^* + \delta) - \delta < 0, \quad \forall 0 < \delta < s, \tag{B.21}$$

(b)

$$\nu_i^* \left[G'(p(\nu_i^* - \delta, \nu_{-i}^*)) - G'(p^*)\right](\nu_i^* - \delta) + \delta > 0, \quad \forall 0 < \delta < \nu_i^*, \tag{B.22}$$

where $s = \max\{a_1, \ldots, a_n\} > 0$. From the proof of Theorem 3.2 [1] we have:

$$0 \leq \nu_i^* \leq \frac{s}{n-2}, \quad i = 1, \ldots, n. \tag{B.23}$$

Now, from assumption (B.19) and the fact that $\frac{\partial p}{\partial \nu_i} > 0$, it follows that

$$\begin{aligned} &\nu_i^* \left[G'(p(\nu_i^* + \delta, \nu_{-i}^*)) - G'(p^*)\right](\nu_i^* + \delta) - \delta \\ &\leq \nu_i^* \left|G'(p(\nu_i^* + \delta, \nu_{-i}^*)) - G'(p^*)\right|(\nu_i^* + \delta) - \delta \\ &\leq \nu_i^*(\nu_i^* + \delta)\frac{1}{2s^2 G(p_0)}|p(\nu_i^* + \delta, \nu_{-i}^*) - p^*| - \delta \\ &\leq \nu_i^*(\nu_i^* + \delta)\frac{1}{2s^2 G(p_0)}(p(\nu_i^* + \delta, \nu_{-i}^*) - p^*) - \delta. \end{aligned} \tag{B.24}$$

By the mean value theorem here exists a value $\hat{\nu}_i$ such that $\nu_i < \hat{\nu}_i < \nu_i + \delta$ and

$$p(\nu_i^* + \delta, \nu_{-i}^*) - p^* = \delta \frac{\partial p}{\partial \nu_i}(\hat{\nu}_i, \nu_{-i}^*). \tag{B.25}$$

Using (3.10) we get

$$\begin{aligned} \frac{\partial p}{\partial \nu_i}(\hat{\nu}_i, \nu_{-i}^*) &= \frac{\dfrac{q_i(p(\hat{\nu}_i, \nu_{-i}^*), (\hat{\nu}_i, \nu_{-i}^*))}{\hat{\nu}_i + a_i}}{\sum\limits_{\substack{k=1 \\ k \neq i}}^{n} \dfrac{1}{\nu_k^* + a_k} + \dfrac{1}{\hat{\nu}_i + a_i} - G'(p(\hat{\nu}_i, \nu_{-i}^*))} \leq \frac{\dfrac{q_i(p(\hat{\nu}_i, \nu_{-i}^*), (\hat{\nu}_i, \nu_{-i}^*))}{\hat{\nu}_i + a_i}}{\sum\limits_{\substack{k=1 \\ k \neq i}}^{n} \dfrac{1}{\nu_k^* + a_k} + \dfrac{1}{\hat{\nu}_i + a_i}} \\ &\leq \frac{\dfrac{q_i(p(\hat{\nu}_i, \nu_{-i}^*), (\hat{\nu}_i, \nu_{-i}^*))}{\hat{\nu}_i + a_i}}{\dfrac{1}{\hat{\nu}_i + a_i}} = q_i(p(\hat{\nu}_i, \nu_{-i}^*), (\hat{\nu}_i, \nu_{-i}^*)) \leq G(p_0). \end{aligned} \tag{B.26}$$

Applying (B.25) and (B.26) to (B.24) we find:

$$\begin{aligned} \nu_i^* \left[G'(p(\nu_i^* + \delta, \nu_{-i}^*)) - G'(p^*) \right] (\nu_i^* + \delta) - \delta \leq & \nu_i^* (\nu_i^* + \delta) \frac{1}{2s^2} \delta - \delta \\ = & \left[\nu_i^* (\nu_i^* + \delta) \frac{1}{2s^2} - 1 \right] \delta, \end{aligned} \tag{B.27}$$

moreover, since $0 < \nu_i^* \leq s$ and $0 < \delta < s$, it follows that

$$\nu_i^* \left[G'(p(\nu_i^* + \delta, \nu_{-i}^*)) - G'(p^*) \right] (\nu_i^* + \delta) - \delta \leq \left[\nu_i^* (\nu_i^* + \delta) \frac{1}{2s^2} - 1 \right] \delta < 0, \tag{B.28}$$

which proves (a).

Analogous to the previous case, we can find that

$$\nu_i^* \left[G'(p^*) - G'(p(\nu_i^* - \delta, \nu_{-i}^*)) \right] (\nu_i^* - \delta) - \delta \leq \left[\nu_i^* (\nu_i^* - \delta) \frac{1}{2s^2} - 1 \right] \delta, \tag{B.29}$$

and, since $0 < \delta < \nu_i^* \leq s$, we have

$$\nu_i^* \left[G'(p^*) - G'(p(\nu_i^* - \delta, \nu_{-i}^*)) \right] (\nu_i^* - \delta) - \delta \leq \left[\nu_i^* (\nu_i^* - \delta) \frac{1}{2s^2} - 1 \right] \delta < 0, \tag{B.30}$$

then

$$\nu_i^* \left[G'(p(\nu_i^* - \delta, \nu_{-i}^*) - G'(p^*)) \right] (\nu_i^* - \delta) + \delta > 0. \tag{B.31}$$

Therefore, the vector ν^* is Nash equilibrium for $n \geq 3$.

Finally, let $n = 2$. We can repeat the steps for the case $n \geq 3$ to get the following inequality:

$$\begin{aligned} & \nu_i^* \left[G'(p(\nu_i^* + \delta, \nu_{-i}^*)) - G'(p^*) \right] (\nu_i^* + \delta) - \delta \\ & \leq \left[\nu_i^* (\nu_i^* + \delta) \frac{2}{\left(\frac{a_1 + a_2}{\varepsilon \min\{a_1, a_2\}} + 3 \max\{a_1, a_2\} \right)^2} - 1 \right] \delta. \end{aligned} \tag{B.32}$$

From

$$\begin{aligned} \nu_i \leq & \frac{1}{2} \left(\frac{a_1 + a_2}{\varepsilon \min\{a_1, a_2\}} + \max\{a_1, a_2\} \right) + \max\{a_1, a_2\} \\ = & \frac{1}{2} \left(\frac{a_1 + a_2}{\varepsilon \min\{a_1, a_2\}} + 3 \max\{a_1, a_2\} \right), \quad i = 1, 2. \end{aligned} \tag{B.33}$$

we have

$$0 < \nu_i^* \leq \frac{1}{2} \left(\frac{a_1 + a_2}{\varepsilon \min\{a_1, a_2\}} + 3 \max\{a_1, a_2\} \right) \tag{B.34}$$

and

$$0 < \delta < \max\{a_1, a_2\} < \frac{1}{2}\left(\frac{a_1 + a_2}{\varepsilon \min\{a_1, a_2\}} + 3\max\{a_1, a_2\}\right), \tag{B.35}$$

thus

$$\begin{aligned} & v_i^* \left[G'(p(v_i^* + \delta, v_{-i}^*)) - G'(p^*)\right](v_i^* + \delta) - \delta \\ & \leq \left[v_i^*(v_i^* + \delta) \frac{2}{\left(\frac{a_1 + a_2}{\varepsilon \min\{a_1, a_2\}} + 3\max\{a_1, a_2\}\right)^2} - 1 \right] \delta < 0. \end{aligned} \tag{B.36}$$

which finally proves (a).

Analogously, to prove (b), it is easy to show that

$$\begin{aligned} & v_i^* \left[G'(p(v_i^* - \delta, v_{-i}^*)) - G'(p^*)\right](v_i^* - \delta) + \delta \\ & \geq \left[1 - v_i^*(v_i^* - \delta) \frac{2}{\left(\frac{a_1 + a_2}{\varepsilon \min\{a_1, a_2\}} + 3\max\{a_1, a_2\}\right)^2} \right] \delta > 0. \end{aligned} \tag{B.37}$$

The theorem has been proved ■

Appendix C: Proofs of Results from Chapter 4

Theorem 4.1 *The quadratic programming problem (4.11)–(4.15) is convex and any of its solutions provides the Nash equilibrium for the non-cooperative game (4.6)–(4.10).*

Proof First, in order to prove that the quadratic programming problem (4.11)–(4.15) is convex, we only need to prove that the symmetric matrix associated with the quadratic objective function (4.12) is positive semidefinite, i.e., we need to prove that

$$\sum_{k\in K}\sum_{\ell\in K\setminus\{k\}}\sum_{a\in A}\frac{1}{2}d_a x_a^k x_a^\ell + \sum_{k\in K}\sum_{a\in A} d_a \left(x_a^k\right)^2 \geq 0,\ \forall x \in \mathbb{R}^{M_K}. \tag{C.1}$$

Indeed, let $x \in \mathbb{R}^{M_K}$, then, we have that

$$\begin{aligned}
&\sum_{k\in K}\sum_{\ell\in K\setminus\{k\}}\sum_{a\in A}\frac{1}{2}d_a x_a^k x_a^\ell + \sum_{k\in K}\sum_{a\in A} d_a \left(x_a^k\right)^2 \\
&= \sum_{k\in K}\sum_{a\in A}\frac{1}{2}d_a \left(x_a^k\right)^2 + \sum_{k\in K}\sum_{\ell\in K}\sum_{a\in A}\frac{1}{2}d_a x_a^k x_a^\ell \\
&= \sum_{k\in K}\sum_{a\in A}\frac{1}{2}d_a \left(x_a^k\right)^2 + \sum_{a\in A}\frac{1}{2}d_a\left(\sum_{k\in K}\sum_{\ell\in K} x_a^k x_a^\ell\right) \\
&= \sum_{k\in K}\sum_{a\in A}\frac{1}{2}d_a \left(x_a^k\right)^2 + \sum_{a\in A}\frac{1}{2}d_a\left(\sum_{k\in K} x_a^k\right)^2.
\end{aligned} \tag{C.2}$$

Since all of the congestion coefficients d_a are nonnegative, we can easily see that (C.2) is also nonnegative. Moreover, if all congestion coefficients d_a are strictly positive, then, (C.2) is also strictly positive and the quadratic programming problem (4.11)–(4.15) is strictly convex so it has a unique solution.

J. G. Flores Muñiz et al., *Public Interest and Private Enterprize: New Developments*, Lecture Notes in Networks and Systems 138,
https://doi.org/10.1007/978-3-030-58349-1

Now, we will prove that any solution of (4.11)–(4.15) provides the Nash equilibrium for (4.6)–(4.10). In order to do this, we rewrite both problems in their matrix forms.

Let $\{t_a \mid a \in A_1\}$ satisfy (4.4) and (4.5), then, we can consider the vector $z \in R^M$ whose a-th component is given by c_a if $a \in A_2$ and by $t_a + c_a$ if $a \in A_1$. Thus, the Nash equilibrium problem (4.6)–(4.10) is given as follows:

$$x^k \in \Psi_k(t, x^{-k}), \ \forall k \in K, \tag{C.3}$$

where

$$\Psi_k(t, x^{-k}) = \underset{x^k}{\text{Argmin}} \ f_k(x^k) = z^T x^k + \sum_{\ell \in K \setminus \{k\}} {x^k}^T H x^\ell + {x^k}^T H x^k, \tag{C.4}$$

$$\text{subject to} \qquad Bx^k = b^k, \tag{C.5}$$

$$x^k \leq q - \sum_{\ell \in K \setminus \{k\}} x^\ell, \tag{C.6}$$

$$x^k \geq 0. \tag{C.7}$$

Here, the matrix H is the diagonal matrix $M \times M$ matrix corresponding to the congestion factors, i.e., the a-th diagonal element of H is d_a. The matrix $B \in \mathbb{R}^{\eta \times M}$ and the vector $b^k \in \mathbb{R}^\eta$ corresponds to the equality constraints (4.8) (the matrix B depends solely upon the network so it is the same for any commodity k), and the vector $q \in \mathbb{R}^M$ has the capacity upper bounds q_a, $a \in A$, as its components. Using the above notation, the quadratic programming problem (4.11)–(4.15) is given by:

$$x \in \Psi(t), \tag{C.8}$$

where

$$\Psi(t) = \underset{x}{\text{Argmin}} \ f(x) = \sum_{k \in K} z^T x^k + \sum_{k \in K} \sum_{\ell \in K \setminus \{k\}} \frac{1}{2} {x^k}^T H x^\ell + \sum_{k \in K} {x^k}^T H x^k, \tag{C.9}$$

$$\text{subject to} \qquad Bx^k = b^k, \ \forall k \in K, \tag{C.10}$$

$$\sum_{\ell \in K} x^\ell \leq q, \tag{C.11}$$

$$x \geq 0. \tag{C.12}$$

The matrix D is a $M\kappa \times M\kappa$ symmetric block matrix whose $\kappa \times \kappa$ block components are the matrix H in its diagonal blocks and the matrix $\frac{1}{2}H$ in its non-diagonal blocks. Since all the values d_a are nonnegative, both, the matrix H and the matrix D are symmetric and positive semi-definite, and if all the values d_a are strictly positive, the matrices H and D will be positive definite.

The programs appearing in (C.3)–(C.7) and program (C.8)–(C.12) are differentiable and convex (with linear constraints) quadratic programming problems, so these problems can be equivalently transformed into a nonlinear system of equations and inequalities using the Lagrange multipliers and Karush-Kuhn-Tucker conditions. Therefore, in order to show that a solution of (C.8)–(C.12) generates Nash-equilibrium for (C.3)–(C.7), it will suffice to demonstrate that a solution for the Lagrange multipliers and Karush-Kuhn-Tucker conditions of (C.3)–(C.7) lead to a solution for the Lagrange multipliers and Karush-Kuhn-Tucker conditions of (C.3)–(C.7). The Lagrange multipliers and Karush-Kuhn-Tucker condition for problem (C.3)–(C.7) are as follows:

$$\frac{df_k}{dx^k} + \mu^k + B^T\lambda^k = z + \sum_{\ell\in K\setminus\{k\}} Hx^\ell + 2Hx^k + \mu^k + B^T\lambda^k \geq 0, \quad \text{(C.13)}$$

$$x^k\left(z + \sum_{\ell\in K\setminus\{k\}} Hx^\ell + 2Hx^k + \mu^k + B^T\lambda^k\right) = 0, \quad \text{(C.14)}$$

$$Bx^k = b^k, \quad \text{(C.15)}$$

$$x^k \leq q - \sum_{\ell\in K\setminus\{k\}} x^\ell, \quad \text{(C.16)}$$

$$\mu^k\left(\sum_{\ell\in K} x^\ell - q\right) = 0, \quad \text{(C.17)}$$

$$x^k, \mu^k \geq 0, \quad \text{(C.18)}$$

where $\mu^k \in \mathbb{R}^M$ and $\lambda^k \in \mathbb{R}^\eta$, for all $k \in K$. And the Lagrange multipliers and Karush-Kuhn-Tucker conditions for problem (C.8)–(C.12) are:

$$\frac{\partial f}{\partial x^k} + \mu + B^T\lambda^k = z + \sum_{\ell\in K\setminus\{k\}} Hx^\ell + 2Hx^k + \mu + B^T\lambda^k \geq 0, \ \forall k \in K, \quad \text{(C.19)}$$

$$x^k\left(z + \sum_{\ell\in K\setminus\{k\}} Hx^\ell + 2Hx^k + \mu + B^T\lambda^k\right) = 0, \ \forall k \in K, \quad \text{(C.20)}$$

$$Bx^k = b^k, \ \forall k \in K, \quad \text{(C.21)}$$

$$\sum_{\ell\in K} x^\ell \leq q, \quad \text{(C.22)}$$

$$\mu\left(\sum_{\ell\in K} x^\ell - q\right) = 0, \quad \text{(C.23)}$$

$$x, \mu \geq 0, \quad \text{(C.24)}$$

where $\mu \in \mathbb{R}^M$ and $\lambda^k \in \mathbb{R}^\eta, k \in K$. Now let x^k, $\mu \in \mathbb{R}^M$ and $\lambda^k \in \mathbb{R}^\eta, k \in K$, satisfy (C.19)–(C.24). Then, for a fixed $k \in K$, we have that:

$$z + \sum_{\ell \in K \setminus \{k\}} Hx^{\ell} + 2Hx^k + \mu + B^T \lambda^k \geq 0, \tag{C.25}$$

$$x^k \left(z + \sum_{\ell \in K \setminus \{k\}} Hx^{\ell} + 2Hx^k + \mu + B^T \lambda^k \right) = 0, \tag{C.26}$$

$$Bx^k = b^k, \tag{C.27}$$

$$x^k \leq q - \sum_{\ell \in K \setminus \{k\}} x^{\ell}, \tag{C.28}$$

$$\mu \left(\sum_{\ell \in K} x^{\ell} - q \right) = 0, \tag{C.29}$$

$$x^k, \mu \geq 0. \tag{C.30}$$

Therefore, the vectors $x^k, \mu \in \mathbb{R}^M$ and $\lambda^k \in \mathbb{R}^{\eta}$ satisfy (C.13)–(C.18), for all $k \in K$, which proves the theorem.

Finally, if we remove the capacity constraints, we can easily see that the KKT conditions (C.13)–(C.18) imply the KKT conditions (C.19)–(C.24) taking $\mu := \max\{\mu^k | k \in K\}$, which proves Corollary 4.1 ■

Reference

1. Bulavsky, V.A.: Structure of demand and equilibrium in a model of oligopoly. Econ. Math. Methods (Ekonomika i Matematicheskie Metody) **33**, 112–134 (1997). In Russian

Printed in the United States
By Bookmasters